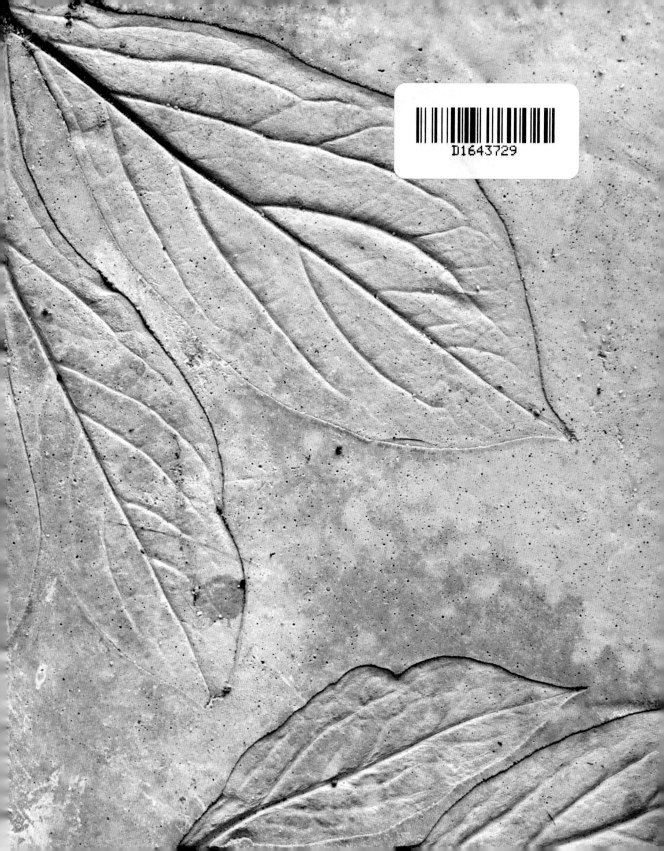

EASY

CONCRETE

projects for home and garden

Published in 2010 by New Holland Publishers (UK) Ltd
London • Cape Town • Sydney • Auckland

www.newhollandpublishers.com

Garfield House, 86–88 Edgware Road, London W2 2EA,
United Kingdom

80 McKenzie Street, Cape Town 8001, South Africa

Unit 1, 66 Gibbes Street, Chatswood, NSW 2067, Australia

218 Lake Road, Northcote, Auckland, New Zealand

First published by Prisma, Sweden, in 2009 as *Inspirerande Betong*
Published by agreement with Norstedts Agency
© 2009 Malena Skote and Prisma, Stockholm 2009
© 2010 English Translation, New Holland Publishers

A catalogue record for this book is available from the British Library

ISBN 978 184773 667 3

Publishing Director: Rosemary Wilkinson
Publisher: Clare Sayer
Senior Editor: Emma Pattison
Photography: Malena Skote, unless otherwise specified
Design: Malena Skote
Illustrations: Malena Skote
Translation: Brett Jocelyn Epstein
Production: Laurence Poos

10 9 8 7 6 5 4 3 2 1

Reproduction by Pica Digital PTE Ltd, Singapore
Printed and bound in India by Replika Press Pvt. Ltd

EASY
CONCRETE
projects for home and garden
Malena Skote

Contents

Foreword

In the mid-1990s, I decided to begin casting concrete pots at home in my garden. I'd read a description of how you could put a smaller plastic pot inside a bigger one and then pour concrete into the cavity between the two. I had to try it!

I planned my purchases carefully, read handbooks, measured the diameters of the plastic pots and set aside a whole day for this new, exciting project. Nowadays, I can run out to the garage on impulse 20 minutes before dinner, mix a small portion of concrete and cast a pot or a plate.

What's great about concrete is that it doesn't require much in the way of preparation. It is a cheap and unpretentious material that is simple to work with. Casting a concrete pot isn't any harder than baking a sponge cake from scratch – you don't even need an oven.

How did it turn out, that first pot? Well, it's still standing there, adorning the steps. It's raw, simple and robust, and lets the plants speak for themselves. It doesn't freeze and break, and it never gets blown over. It has confirmed my initial thoughts about using concrete – it's grey, rough, robust and provides a frost-resistant background for my plants.

I soon realized just how versatile concrete can be, how many pieces you can make from it (not just strong pots), and how different kinds of concrete can look. And it doesn't have to be grey and rough – it can be colourful with a touch of added pigment, or smooth and fine when cast against something like an old silver tray. Since that first pot I have made many experimental pieces with concrete, and tried countless ideas and new materials, but above all, I've played and had fun.

If you think something is really enjoyable you want to share it with others, so in this book I've gathered together over ten years worth of tips and ideas from my own experiments and from courses and workshops. Thank you to all those who have contributed over the years!

Malena Skote

My first pot to the left and newer pots to the right.

About moulds

It isn't possible to write about concrete without writing about moulds, since they determine how the finished objects will look. You pour concrete into a mould like batter into a cake tin. The concrete hardens in the mould and the finished object takes on the shape of the inside of the mould. This means that when you make moulds for concrete casting you have to imagine the negative image – that is to say, the cavity into which the concrete is poured.

The material from which the mould is made also affects how the concrete object will look. The surface of the concrete mirrors the surface it is cast against – it will be shiny and smooth if it is cast against a piece of glass, and matt and rough if it is cast against sand.

Inspect the mould carefully. What material is it made of? Is the material smooth or matt? Are there trademark symbols that might make an imprint on the concrete? Don't forget that even price tags left on the surface can leave impressions! And remember that the mould has to be strong enough so that it doesn't collapse when you pour in the wet concrete.

Concrete is a somewhat surprising material. When you pour it into the mould you have no control over what will happen. You simply have to be patient and await the exciting moment when you can see your brand new concrete object.

Moulds made from juice cartons. The cartons have waxed insides that easily release the cement but must be supported from the outside with bricks so they don't fold over when the heavy concrete is poured in.

Concrete bee hives by Hillevi Strandlund, cast in plastic biscuit tins.

A sculpture by Nilgün Fernehall cast in a plastic bowl. The bowl shape was made by lowering a smaller container into wet concrete. When the concrete hardened, the bowl was covered in glass mosaic tiles and a mirror was placed at the bottom.

Mushroom footstools by Nilgün Fernehall. The caps were cast in washtubs and the stems in plastic buckets.

Plastic moulds

Plastic pails, pots and washtubs are available cheaply at flea markets and large stores. Save ice cream, biscuit and cream containers, and ask for empty candy boxes at stores.

When you cast pots or bowls, you usually place a smaller bowl or plastic pot in a bigger one and pour the concrete into the cavity in between the two. So you need to try to imagine the volume of the space in between. How will the proportions turn out? How thick will the walls of the bowl/pot be? The outer (larger) mould's inside should be as smooth as possible. Are there measurement markings that can leave impressions? For the inner (smaller) mould, it is the outside that should be smooth so that it doesn't get stuck to the concrete. Saw or file away rough edges and tape over draining holes in plastic pots.

And remember that if the mould is flexible and the plastic is somewhat smooth, you won't need to oil it.

Wooden moulds

Wooden boxes, cases and plates are available at flea markets and large stores or you can easily make them yourself from pieces of wood. If the wood is untreated and seems absorbent lacquer it first, otherwise the wood will absorb the water from the concrete and the concrete surface may turn sandy. Use transparent wood or acrylic lacquer. Formwork plywood is especially made for concrete casting and is covered with a film that makes it release the concrete easily. It is sold in large pieces of 2.4 x 1.2 metres (7 ft 10 in x 4 ft) in building materials stores. If you don't want such a big piece, you can ask at construction sites for leftovers. Used formwork plywood can usually be scraped clean and re-used. With wooden moulds, you get straight corners and sharp edges on your concrete objects, in contrast to the rounded edges you get from plastic moulds. A wooden mould must be tight and without gaps, or else the water in the wet concrete will run out and the edges will be crumbly. Screw your mould together, instead of nailing, so the joints will be tighter; it will also be easier to take the mould apart. Formwork plywood is easy to screw into if you first make holes with a sharp awl.

Concrete tiles being made in small wooden moulds.

Metal
Although you can make concrete plates on top of metal ones (see pages 53–57), for deeper moulds, such as cake tins and saucepans that are rigid and cannot be stretched, the concrete can get stuck. Oiling the mould first with a thin layer of lubricant will help prevent the concrete from adhering to the sides. Avoid using tins because the ridges will hold the concrete fast.

Concrete plate cast using a silver-plated dish (see pages 53–57).

Paper
Ordinary, flimsy cardboard will dissolve in wet concrete, but firm paper, such as paper tubes from building materials stores and boxes can be used as moulds. Milk and juice cartons have waxed interiors that easily release concrete, but they must be supported from the outside so they don't bow and split when the heavy, wet concrete is poured in.

Concrete shells cast in soap moulds from the craft store.

Silicone and rubber

Mould silicone and liquid rubber can be used to make moulds for more complicated figures, such as animals and fruits. Choose a good original, such as a beautiful apple, cover it in silicone or rubber and let it harden, then remove the apple and pour concrete into the mould. Mould silicone and rubber are available in craft stores, art stores and shops that sell materials for ceramics. Pre-made moulds for plaster, soap and so on can be used for concrete, and they are available in craft shops.

Lubricant spray creates a thin layer of oil in the mould.

Release agent

Greasing the mould with lubricant makes it easier to release the finished concrete piece. Cooking oil and petroleum jelly can be used, but they will make the surface of the concrete very porous. A porous surface can be beautiful if that is the effect you are after, but if you want a completely smooth surface, use a thin liquid oil specifically for this purpose (mould oil is available from building materials stores). Lubricant sprays, such as those used by bakers or the type meant for mechanical machine parts, can also be used.

Using concrete

Small pots made from different types of concrete. The little one towards the front of the picture is made from 'flow concrete' which creates a smooth, poreless surface.

oncrete is made of cement and sand and cement is, in turn, made of clay and ground limestone. When the cement-sand mixture is combined with water the cement reacts by hardening and creates a hard 'glue' between the grains of sand.

There are many different dry cement products available from building materials stores, where the sand has already been mixed in and you simply add water. However there are several main groups in the cement 'jungle' and it is worth knowing them and being able to differentiate between them.

Types of concrete

- **Fine concrete** An all-round concrete that can be used for most craft items. Sold in large sacks, it contains grains of sand that are no larger than 4 mm (⅙ in). Frost-resistant.

- **Coarse concrete** Contains coarser-grained sand. Works best for stairs, plinths, etc. Frost-resistant.

- **'Flow concrete', expander concrete, floor-levelling concrete** Concrete with fine grains of sand that swells and fills small and intricate moulds particularly well. Creates a smooth surface, almost without pores. Often not frost-resistant.

Using a plastic putty knife you can shape and sculpt your concrete objects.

You should be able to shape the mixture into a ball.

The ball should flow out when you shake your hand.

- **Mending concrete, concrete putty, levelling putty** Pliable, plastic and mouldable cement-based mortar. Doesn't flow out and doesn't run over in general. Sold in cartons and medium-sized sacks. Can be found in a frost-resistant form.
- **Fibre concrete** Contains small plastic or glass bits to prevent cracks when casting thin surfaces such as plates and tabletops. Has fine sand and is frost-resistant.

Mixing tips

Using a strong plastic pail or tub, pour in a little water to prevent the powder sticking to the bottom. Wearing gloves to protect your hands scoop a couple of ladles of concrete powder into the pail (avoid breathing in the concrete dust). Add a little water at a time and stir until the mixture is thick and viscous. Do the 'meatball test': you should be able to shape the mixture into a ball, but the ball should collapse when you shake your hand. Half a pail of concrete is about all you can mix by hand – for larger amounts use a cement mixer.

Once you are finished, rinse your tools and pour water into the pail to prevent the concrete from sticking. Do not pour concrete mixed water down the drain. Leave it until the concrete chalk settles to the bottom. The water will clear and can be poured out, and the concrete can be scraped out and thrown away.

Casting and hardening

Fill the mould almost all the way up. When casting bowls and pots, it's a good idea to leave a little 'pinch bit' at the top of the mould, so you can get the object out more easily. Shake or tap the mould so the concrete settles all the way down. Press the concrete down with a trowel or a stick. When you shake the mould you will see small air bubbles coming up through the wet concrete surface. The more air bubbles that come up, the smoother the concrete surface will be, with fewer pores.

In order to make the concrete as strong as possible, it is important that it doesn't dry too quickly. Cover the mould with plastic so the water doesn't evaporate.

Press the concrete down into the mould so that all air pockets disappear.

Unpacking

Of course you will want to see your object as soon as possible, but be patient! Concrete may seem hard after just a few hours, but it continues to harden for several weeks. The hardening process depends on temperature, humidity, the type of concrete and the material of the mould, so it is difficult to say exactly how much time it needs. I usually wait two days before opening the mould, just to be on the safe side. Be extra careful with moulds that have sharp corners and thin edges before the concrete is thoroughly hardened.

Allowing for a 'pinch bit' at the top of your object will help with the un-moulding process.

Storage

Keep your sacks of unused concrete dry. If any moisture gets in the concrete will start to harden, get lumpy and then will have to be thrown away. Any opened sacks of concrete should be placed in closed plastic bags or in pails with tight covers.

Round pot in white cement and trough in hypertufa. Both are made from cement but have completely different textures.

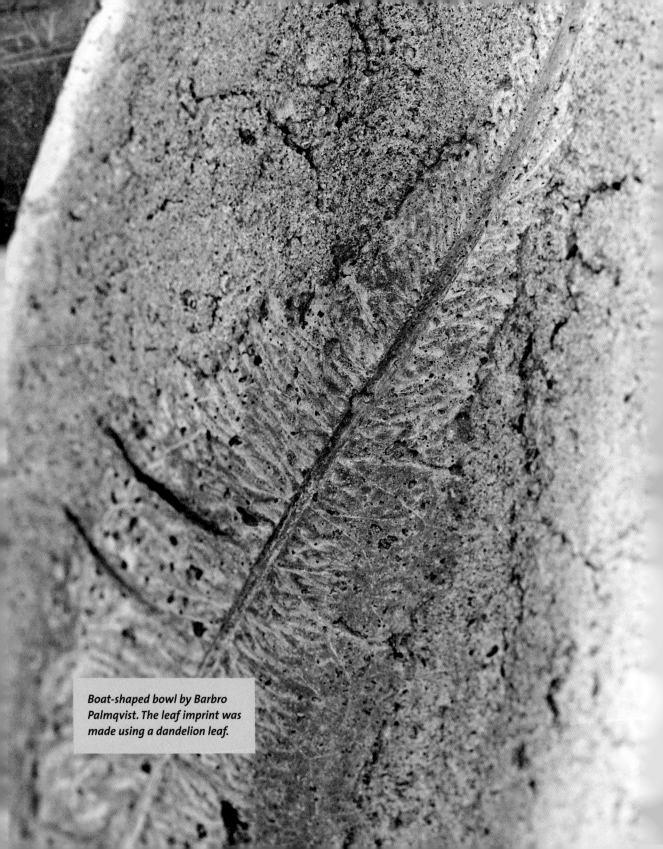

Boat-shaped bowl by Barbro Palmqvist. The leaf imprint was made using a dandelion leaf.

Casting in sand

This mound of sand has been created in order to cast a birdbath.

Wet sand can be shaped by your hands as easily as castles and other short-lived sculptures on the beach. Making moulds out of sand for concrete casting is also cheap, and once you are finished with the mould you simply brush or rinse away the sand.

Concrete surfaces that are cast against sand will be somewhat rough and a little creased, which creates a weathered look. You can cast bowls and plates over mounds of sands and also in depressions of sand. You can use any kind of sand as long as it's wet.

Types of sand

- **Casting sand** for silver, ceramics and glass casting is available from craft shops.
- **Fine sand** for aquariums is available in pet stores.
- **Sandbox sand** can be used although the sand might contain pieces of concrete making it unsuitable for sandbox use.
- **Masonry sand, plaster sand and casting sand** are sold in sacks at building materials stores or can be ordered on the internet. The sand will have different sized grains depending on its intended use.

Mould
Wet sand

Materials
3 corks from wine bottles
Fine concrete
Oxide red pigment (optional)
Rope – 15 mm ($\frac{5}{8}$ in) in diameter and
    approximately 1.5 m (60 in) long

Tools
Mixing pail
Trowel or steel putty knife
Plastic sheet

Instructions
1. Working directly on the ground or on a durable or plastic-covered workspace, make a round mound out of the wet sand and shape a flat edge of about 10 cm (4 in) around it. The mould should look like a hat with a round brim.

2. Place the 3 corks evenly spaced around the flat edge (see illustration, left). This will create the holes for the rope.

3. Mix several litres/quarts of dry concrete with a small amount of water, adding a little at a time until it has the consistency of porridge and is neither crumbly nor runny. Mix in the red pigment, if using (less than a tablespoon is usually enough to turn the concrete red).

4. Wearing rubber gloves shape a 2 cm ($\frac{3}{4}$ in) thick layer of the concrete mixture over the sand mould. Pat it lightly with your hand in order to get rid of any air pockets between the sand and the concrete and also to smooth out the surface. Even out the edges using the trowel or putty knife.

5. Cover with the plastic sheet and leave the birdbath to harden for two days.

6. Turn the birdbath over and brush or rinse away the sand. Carefully push out the corks. Cut the rope into three equal lengths, make sturdy knots at one end of each length and thread the other end of the rope through the holes made by the corks.

A hanging birdbath in red-coloured concrete by Sólveig Fridriksdóttir.

Mould
Wet sand

Materials
Fine concrete
Large dandelion leaf or other
   decoration (optional)
Liquid gold leaf (optional)

Tools
Marker pen
Mixing pail
Trowel
Plastic sheet

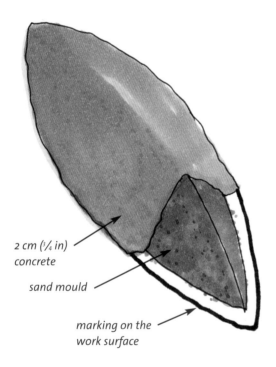

2 cm (³⁄₄ in)
concrete

sand mould

marking on the
work surface

Instructions
1. Working on a piece of wood or on a plastic-covered work surface make a mound out of the wet sand and shape it like an upside-down boat.

2. Using the marker pen draw a line around the mound about 2 cm (¾ in) away from the edge. When the concrete is shaped it should reach this line but not cross it. It may be difficult to achieve an even thickness all the way around the bowl but the line should help.

3. Mix several litres/quarts of dry concrete with a little water until it forms a viscous paste that is neither runny nor crumbly. It should be moist all the way through but still stick together. If you are using the dandelion leaf or other decoration place it diagonally on the sand mound with its back facing outwards.

4. Wearing rubber gloves shape a layer of concrete about 2 cm (¾ in) thick over the sand mound. Even it out as much as possible with your hands or use a trowel. Flatten the surface at the top of the moulded concrete so that the bowl will stand steady when placed upright. Cover with the plastic sheet and leave to harden for two days.

5. Turn the bowl over, brush away the sand and remove the leaf. Paint the stem of the leaf using the liquid gold, if liked.

Boat-shaped bowl by
Barbro Palmqvist.

To make this umbrella stand use the base of a 'lathed' plastic urn. Turn the urn upside-down and press into into moist sand to make an impression of the shape of the urn.

Use a cardboard tube to make the hole with a diameter that is somewhat larger than that of an umbrella. Alternatively cast a metal roll into the umbrella stand.

Mould

Large sturdy cardboard box or plastic tub, large enough to fit the plastic urn stand inside
Wet sand
Plastic urn stand
Cardboard or metal tube, with a diameter that is somewhat larger than that of an umbrella

Materials

Fine concrete

Tools

Mixing pail
Trowel
Plastic sheet

Plastic urn stand

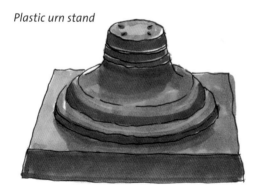

Instructions

1. Fill the cardboard box or plastic tub with wet sand. Push the plastic urn stand down into the wet sand and then carefully pull it out. Place the cardboard or metal tube into the middle of the imprint.

2. Mix several litres/quarts of dry concrete with a little water until it forms a viscous paste, referring to the instructions on the packet. Carefully pour the concrete into the impression, holding the paper/metal tube steady so it doesn't lean sideways. Cover with the plastic sheet and leave the concrete to harden for two days.

3. Remove the concrete stand from the sand. Remove the tube and brush or rinse the sand away.

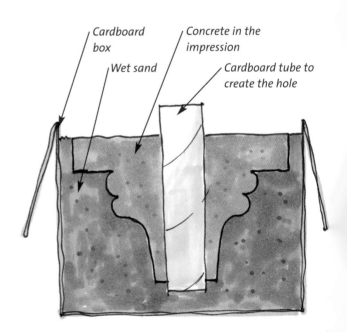

Cardboard box

Concrete in the impression

Wet sand

Cardboard tube to create the hole

Umbrella stand by Kirstina Digman.

This double birdbath was made on top of two mounds of sand. The mosaic pieces were placed around it and a little edge of clay stopped the concrete from running out onto the work surface. In this way, the concrete mixture could be made fairly loose so it would run out between the mosaic pieces. As an experiment, white pigment was scattered directly onto the sand before the concrete was placed on top.

Mould
Wet sand
Mosaic pieces
Pottery, sketch, craft or modelling clay

Materials
Fine concrete

Tools
Mixing pail
Trowel
Plastic sheet

Instructions

1. Working on a piece of wood or directly on a plastic-covered work surface, make two even mounds of sand (see diagram below) and even out the surface. Place the mosaic pieces around the mould, good side down.

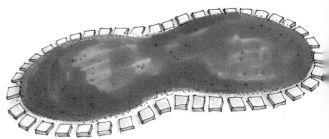

2. Press a border of clay around the outside of the mosaic pieces.

3. Mix several litres/quarts of concrete powder with a little water until it forms a viscous paste. Place a 2 cm (¾ in) thick layer of concrete onto the mounds and the mosaic pieces. Cover with a plastic sheet and leave to harden for two days.

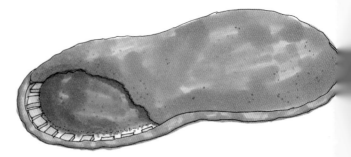

4. Turn the birdbath over and brush or rinse the sand away. If any of the mosaic pieces come loose glue them in place with an adhesive designed for outdoor use.

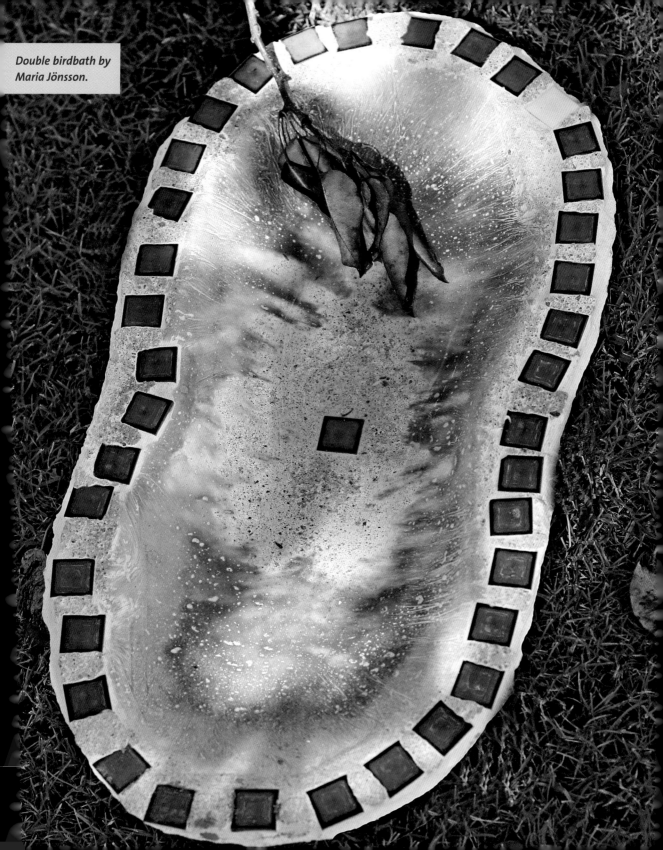

Double birdbath by Maria Jönsson.

Blue pot in fibre concrete by Inga-Lill Lager. A leaf was fastened to the inside of the outer mould with spray glue, which created a beautiful imprint in the concrete.

Blue concrete

Even though I like the natural, warm grey colour of concrete, I think it's fun to experiment with coloured concrete. You can use any pigment to colour concrete, but I have achieved the best results by using liquid pigments. They are expensive, but also last an unbelievably long time. You need to use around a tablespoon of pigment for every half a pail of concrete to make the paste really blue.

Concrete with pigment in it takes longer to harden and isn't as strong, so don't make coloured pots too thin. Use fibre concrete that has in-built reinforcement from the small glass or plastic pieces contained within it.

Mould

Tall plastic plant pot (sold in garden centres) for the outer mould. For the inner mould use a large coffee cup, a tall yogurt pot or a slightly smaller plastic pot. There should be at least 1.5 cm (⅝ in) space between the moulds.

Materials

Packing tape
Fine or fibre concrete
Ultramarine liquid pigment
Cork from a wine bottle
Sand or small stones

Tools

Mixing pail
Trowel
Old tablespoon
Plastic sheet
Hammer
Coarse sandpaper

Instructions

1. Tape over the hole in the larger pot from the outside. Mix a couple of litres/quarts of concrete powder with a little water until it forms a porous, viscous paste. Blend in about 1 tablespoon pigment. Mix thoroughly. Add a little more pigment if you think the colour is too weak or too greenish.

2. Pour about 2 cm (¾ in) concrete into the bottom of the plant pot. Press the cork down into the concrete to create a draining hole in the finished concrete pot.

3. Fill the smaller mould with sand or small stones and place it on top of the cork. If you are using a plastic plant pot as the inner mould tape over the hole in the bottom first. Spoon the concrete down into the cavity between the moulds. Fill the space almost to the top, or as high as you want the finished pot to be. Hold both moulds and shake them so the concrete evens out and air bubbles escape to the surface. The more air bubbles that escape, the less porous the finished surface will be. Cover with the plastic sheet and leave to harden for two days.

4. Turn the pot over, pour out the sand and draw out the inner plastic pot. Stretch the outer pot, turn it upside-down and shake out the concrete pot. If it is stuck, you can pour a little hot water onto the edge so it slips down between the plastic and the concrete.

5. Tap the cork out carefully using a hammer and smooth the edges of the pot with sandpaper.

White cement (see page 37) with blue pigment added makes a pretty baby-blue concrete pot.

White concrete

It is possible to find a type of cement that is almost completely chalk-white when set. The white colour is achieved by using especially clean raw materials and a special burning and grinding technique.

If you mix white cement with white sand or with crushed white marble, you get white concrete. White cement is available in large sacks. They are almost never in stock at building materials stores but you can ask them to order it from their suppliers. Crushed white marble is available at gardening stores and building materials stores.

The white colour inspired me to make small, round half-egg-shaped pots and a large urn, cast in a lampshade. When you work with white concrete, it is important to use completely clean moulds as any dirt will stick to the white cement and ruin the effect you are trying to achieve.

Mould

Large plastic plant pot or bowl for
the outer mould
Smaller plastic plant pot or bowl
for the inner mould

Materials

Sand
Packing tape (if needed)
White concrete
Crushed white marble
Cork from a wine bottle, halved

Tools

Clean mixing pail
Trowel
Old tablespoon
Plastic sheet
Hammer
Coarse sandpaper (optional)

Instructions

1. Check that the moulds are clean. Fill the inner mould with sand – if you are using a plastic plant pot tape over the draining hole from the inside first.

2. Carefully mix one part white cement with two parts crushed white marble. Pour in a little water at a time until it becomes a viscous, grainy paste with the consistency of cottage cheese. Pour about 2 cm (³/₄ in) of the mixture into the bottom of the larger mould and press half a cork into the centre of the concrete to create a draining hole in the bottom of the pot. Place the smaller mould inside the larger one (on top of the cork) and spoon the concrete down into the space between the moulds. Holding the inner mould so it doesn't slip sideways, shake the pots so the air bubbles come to the surface. Cover with a plastic sheet and leave to harden for two days.

3. Turn the moulds over, pour out the sand and draw out the inner pot. Stretch the outer pot, turn it upside-down and shake out the concrete pot. If it doesn't want to come out, try pouring a little hot water onto the edge so it slips down between the plastic and the concrete.

4. Tap the cork out carefully using a hammer and smooth the edges of the pot with coarse sandpaper, if liked.

Half-egg-shaped pots cast in plastic bowls. The stand was made using a tall yogurt container.

This large pleated pot was cast in a lampshade. The thin edges on the pleats got a little chipped, but this helps to give the pot an aged look.

Mould

Plastic tub, such as a washing-up bowl, large enough to fit the lampshade inside
Pleated lampshade

Materials

Sand
Cork from a wine bottle
White cement
Crushed white marble

Tools

Clean mixing pail
Trowel
Hammer

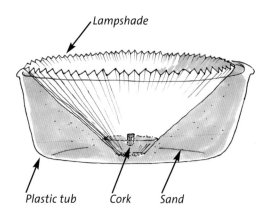

Lampshade

Plastic tub   Cork   Sand

Instructions

1. Fill the plastic tub with sand and make a cone-shaped indentation in the sand. Turn the lampshade upside-down and place it into the sand, checking that the shade is supported by the sand around it. In order to get a flat bottom on the pot, pour a little sand into the bottom of the lampshade and smooth it out. Press the cork into the middle of the sand at the bottom of the lampshade, so that half of the cork sticks up out of the sand.

2. Carefully mix one part white cement with two parts crushed white marble. Add a little water at a time until it forms a viscous, grainy paste with the consistency of cottage cheese.

3. Pour the concrete paste around the cork so that the pot has a bottom that is about 2 cm (¾ in) thick. Pat the concrete down onto the inside of the lampshade, from the bottom upwards. Press with the trowel so that the concrete goes into the pleats and so the inside of the pot is smooth. The paste should be so firm that it doesn't run. If the lampshade is large, you may need to mix more concrete. Don't make the walls of the pot too thin; they should be at least 2 cm (¾ in) thick. Cover with a plastic sheet and leave to harden for two days.

4. Turn the pot upside-down and carefully rip away the lampshade. Brush the sand away and carefully tap out the cork using a hammer.

*Shells of white cement and crushed white marble,
cast in soap and plaster moulds from the craft
shop. To learn more about casting in soap moulds,
see page 142.*

Giant pot

Concrete works well for large garden pots – the pots won't freeze or blow over, and they get more beautiful each year. Large, black pails or tubs, in different sizes, make excellent moulds. There should be 5–10 cm (2–4 in) difference in the diameter of the moulds, so the concrete pot's walls are quite thick.

Mould

A large plastic pail or tub for the outer mould and a smaller one for the inner mould

Materials

Fine or other concrete (for an outer mould with a 60 cm/24 in diameter and an inner one with 50 cm/20 in, you will need about 25 kg/50 lb)
Cork from a wine bottle
Sand

Tools

Mixing pail
Trowel
Old teaspoon
Plastic sheet
Hammer
Coarse sandpaper
Ladle or large spoon

Instructions

1. Mix several litres/quarts of concrete powder with a little water until it formes a porous, viscous paste.

2. Pour in about 4 cm (1½ in) concrete into the bottom of the larger pail. Press the cork into the centre of the concrete to create a draining hole in the finished concrete pot.

3. Fill the smaller pail with sand and place it in the larger pail on top of the cork. Spoon the concrete down into the space between the pails. Fill it up almost to the top, or as high as you want the concrete pot to be. Mix more concrete when the previous batch runs out. Hold both moulds and shake them so the concrete evens out and air bubbles escape to the surface. The more air bubbles that come up to the surface, the less porous the finished concrete surface will be. Cover with the plastic sheet and leave to harden for two days.

4. Scoop out the sand and remove the inner pail. Stretch the outer pail and shake out the concrete pot. If it is stuck, pour hot water onto the edge so it slips down between the plastic and the concrete.

5. Tap the cork out using a hammer. Smooth the edges of the pot with sandpaper.

Large pot cast in a pail with a diameter of 40 cm (16 in). For the inner mould I used a regular plastic pail.

Large concrete bowl cast in a plastic salad bowl. I glued some peony leaves to the inner mould, which was a hanging flower pot, to make an impression in the concrete.

Decorative bowls

Large decorative bowls have always appealed to me. You don't have to fill a bowl with anything; it can be decorative just as it is. The interior and exterior can have different finishes and the edge can be decorated. There is a lot of surface to play with if you are moulding your own large vessels.

Concrete bowls can be cast in a mould made from two plastic bowls. You could use a large plastic salad bowl or a mixing bowl with a rounded inside. You have to make sure there are no measurement markings or other reliefs on the inside that could leave an impression in the concrete surface.

The inner (smaller) mould should have a softly rounded outside without any edges. Plastic hanging flower pots sold in garden centres are perfect. The outside is smooth, they come in at least two sizes and can be used many times.

I made the bowl on the left with an interior leaf motif by pouring concrete into a plastic salad bowl and then pressing down into the cement with a hanging flower pot that had leaves glued to the outside.

Mould made from a plastic hanging flower pot inside a salad bowl, with concrete in between. The red pieces are used as spacers to make sure the inner mould doesn't knock into the sides of the exterior mould.

Mould

Large plastic salad bowl or other plastic dish as the outer mould and a smaller plastic hanging flower pot as the inner mould. There should be a 2 cm (¾ in) gap between the smaller and larger moulds.

Materials

Peony leaves
Spray glue
Cork from a wine bottle, halved
Sand, stones or other weights
Strip of foam
Double-sided tape
Fine or other concrete
Stone wax, stone soap or paraffin oil

Tools

Knife
Ruler and marker pen
Measuring jug
Mixing pail
Trowel
Plastic sheet

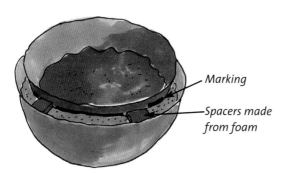

Marking

Spacers made from foam

Instructions

1. Cut away the edge of the hanging flower pot. Spray-glue the peony leaves to the outside of the pot with their backs outwards, and flatten them out as much as possible.

2. Place half of a cork into the centre of the large plastic bowl then place the flower pot, with a little sand or stones in it, on top of the cork and measure the space between the two moulds. Cut pieces of foam to fit that measurement and use tape to fasten them around the flower pot, so that the inner mould doesn't slip sideways causing the pot to have an uneven edge. Also mark on the flower pot the height of the concrete bowl's edge.

3. Measure the concrete bowl's volume by pouring water into the cavity between the moulds and checking to see how many litres/quarts it will hold – this will tell you how much concrete to use – then remove the flower pot and cork.

4. Mix the concrete powder with water into a viscous paste and pour in the same amount as the volume of water you measured.

5. Press the flower pot down until the concrete flows up to the mark. If the flower pot starts floating upwards, add more weight. Check that the spacers are keeping the flower pot in the middle. Cover with a plastic sheet and leave to harden for two days.

6. Turn the pot over, pour out the sand and draw out the flower pot. Stretch the salad bowl so that air gets between the plastic and the concrete then turn the mould upside-down and shake out the concrete pot. If the mould is stuck, you can pour a little hot water onto the edge so that it slips between the plastic and the concrete. If the mould still won't move you might have to cut into the plastic bowl and break it apart in pieces.

7. Treat the bowl with stone wax, stone soap or paraffin oil, so that it can resist dirt and oils.

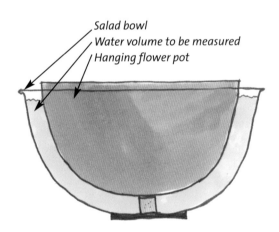

Salad bowl
Water volume to be measured
Hanging flower pot

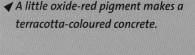

► A smooth bowl in fibre concrete, cast in two plastic salad bowls.

◄ A little oxide-red pigment makes a terracotta-coloured concrete.

◄ The wide edge of this bowl is decorated with leaf motifs.

▲ This bowl's porous concrete surface was created using cooking oil.

▼ A low, wide bowl with a draining hole is good for potting house plants.

Plates

Sisters Ann and Louise Turlock have been casting concrete in their spare time for a number of years. They now specialize in casting concrete plates from the tops of old silver-plated plates. The concrete plates become copies of the metal plates, with patterned borders and filigree. Ann and Louise don't use oil or any other release agent, but place the concrete directly on the tray's back. They then glaze the finished concrete surface white and the result is slender, almost ceramic-like plates.

There are plenty of old silver-plated plates and gold-plated trays available at flea markets and second-hand stores. Study how the back of the tray looks, since it is this side that will create the impression in the cement. Check also that the edge isn't curled back as this can cause the tray to get stuck in the concrete. The tray should be somewhat flexible so that it's easy to extract from the concrete. You can make the plates quite thin but must be as careful with them as you would be with ceramic plates.

A pile of silver plates from the flea market. They make excellent moulds to use for concrete plates.

◄ Plate cast on top of a wooden plate covered with a crocheted cloth.

► Ann and Louise decorating their concrete plates.

◄ Plate cast on an upside-down wooden plate covered with a crocheted cloth.

▼ Plate with silver pattern created using a plastic crocheted cloth as a stencil.

◄ Plate with white border painted after glazing.

Mould

Silver plate with ornamental border

Materials

Fine concrete

White, water-based paint

Small furniture pads

Tools

Mixing pail

Trowel

Metal spatula

Plastic sheet

File

Thin, tapered metal knife

Paintbrush

Rag

Instructions

1. Mix a couple of litres/quarts of concrete powder with a little water at a time until it becomes spongy and sticks together. Hold the plate upside-down and add a little concrete. Shake it with your hand so the concrete flows around. Add more concrete and shake, and continue in the same manner until the plate's back is covered with about 1.5–2 cm ($^5/_8$–$^3/_4$ in) concrete, depending on the size – the larger the plate, the thicker the concrete layer should be.

2. Carefully put the plate down and even out the edges with a metal spatula so the concrete just covers the edges but doesn't run over. Cover with a plastic sheet and leave to harden until the following day.

3. Lift up the plate and file around the edge using the metal plate as a support, then leave to harden for another day before removing the mould.

4. Press the edge of the metal knife between the metal and concrete plates. Do the same all the way around the plate until the vacuum between the metal and concrete loosens. Leave the concrete plate to dry for a few more hours.

5. Dilute a little white paint with water so it is thin and runny, like milk. The thicker the paint, the whiter the end result will be. Paint the whole plate in one go and immediately dry it off with a rag. The pigment usually gets stuck in folds and gaps, which makes the pattern on the border stand out more. If the glazing turns out to be too weak, just paint on another layer.

6. Place small, adhesive furniture pads on the bottom of the plate to prevent it scratching the surfaces of furniture.

Glazing tips

- If you are not happy with the first glaze, simply rinse off the paint. Most of the glaze will disappear and you can start over.
- White-glazing with water-based paint is not weather-resistant. The colour usually disappears after a summer outdoors. Keep white-glazed plates inside or at least under a roof, so they remain beautiful.

A wonderful silver-plated bowl used as a mould for concrete. The embossed pattern on the edges is visible even on the back and creates a faint, barely perceptible pattern in the concrete.

CONCRETE PLATE CAST ON A SMOOTH WOODEN PLATE

Mould
Round wooden plate or tray with a smooth back (if the wood seems absorbent, lacquer it before use)

Materials
Wood or craft lacquer (if needed)
Fine concrete
White, water-based paint
Small furniture pads

Tools
Marking pen
Mixing pail
Trowel
Spatula
Plastic sheet
Thin metal knife
Paintbrush

Instructions
1. Lacquer the plate or tray, if needed. Place the plate upside-down on a durable work surface. Draw a circle around the plate, about 2 cm (¾ in) from the edge, or as wide as you want the edge of the plate to extend.

2. Mix a couple of litres/quarts of concrete powder with a little water until it becomes spongy and sticks together. Pour about 2 cm (¾ in) onto the plate and to the marked edge. Smooth the concrete out using a trowel and even out the edges with a spatula. Cover with a plastic sheet and leave to harden for two days.

3. Press the edge of a thin metal knife between the wooden plate and the concrete. Continue all around the plate until the concrete loosens from the mould. Leave the plate to dry for a few more hours. Since it is hard to shake all the air bubbles from such a mould, the plate's surface will be somewhat porous and creased.

4. Dilute a small amount of white paint with water so it's thin and runny. Paint the plate in one go and immediately dry it off with a rag. The pigment sticks in folds and gaps, which makes the pattern on the border strong. If the glazing is too weak, just paint on another layer.

5. Place small, adhesive furniture pads on the bottom of the plate so it doesn't scratch the table surface.

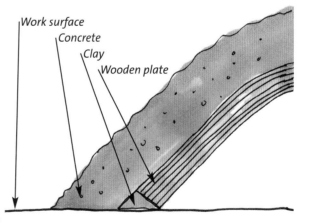

Work surface
Concrete
Clay
Wooden plate

If there is a gap between the edge of the upside-down plate and the work surface, fill it with modelling or craft clay to prevent the concrete from creeping into the cavity, making it difficult to loosen and remove the wooden plate.

*Plate with a mirror mosaic fish
motif by Anna-Lena Ekenryd.*

A shimmering fish in mirror mosaic adorns this little plate, which was cast using a round, plastic plate. The concrete was coloured dark-gray using oxide black pigment.

Mould
Round plastic plate

Materials
Broken mirror
Double-sided tape
Fine concrete
Oxide black pigment (optional)

Tools
Rag or newspaper
Hammer
Mosaic cutters
Scissors
Mixing pail
Trowel
Plastic sheet
Metal spatula or thin knife
Coarse sandpaper

Instructions

1. Wrap the mirror in a rag or newspaper and break into small pieces using a hammer.

2. Cut the mirror pieces into the desired shapes using the mosaic cutters, holding a gloved or otherwise protected hand around the cutters to prevent the pieces from scattering away.

3. Place the mirror pieces onto the tape with the mirror surface downwards and then cut around them so the tape covers the whole mirror surface but doesn't stick out beyond it (if it does, it will leave an impression in the concrete). Stick the pieces to the back of the plastic plate.

4. Mix about ½ litre (½ quart) of concrete powder with a little water until it becomes spongy and sticks together. Add a little of the pigment, if liked. Pour around 1.5 cm (⅝ in) concrete over the whole plate and mirror mosaic. Smooth it out using the trowel ensuring that the concrete doesn't run out over the edge of the plastic plate. Cover with the plastic sheet and leave to harden for two days.

5. Press the edge of a metal spatula or thin knife between the plastic plate and the concrete. Continue all around the plate until the concrete loosens from the mould. Remove the tape from the mirror mosaic and smooth the edges using coarse sandpaper.

Porcelain nostalgia

S tig Lindberg's Berså tableware from the 1960's must be one of Sweden's most beloved tableware designs. Almost everyone loves the pattern with its stylized vines. So do I, but I have never been lucky enough to own a whole cup or plate. However Stig's designs inspired me to experiment with broken porcelain as a decoration for my own concrete pieces. A single shard can decorate an entire pot and many shards can be used to create completely new patterns. Broken porcelain has little value among collectors, so antique shops and flea markets sometimes discard chipped or damaged pieces.

I came accross some rectangular popcorn containers made from hard plastic in an interior design store. They came in different sizes and made great moulds for tall, straight concrete pots. Once you start casting in concrete, you see items in stores with completely different eyes. You see volumes, cavities and negative shapes where you previously saw plastic pots, freezer containers and storage boxes.

LEFT: Nostalgia pots with shards from Berså (by Stig Lindberg) and Mon Amie (by Marianne Westman) cast in hard plastic popcorn moulds.

Mould

Tall rectangular plastic containers (such as popcorn boxes, see page 63) in two different sizes. There should be a cavity of about 1–2 cm (³/₈–³/₄ in) between the two moulds. Cut away any protruding edges from the bottom of the inner mould before use to prevent the mould from sticking into the concrete, making it difficult to release.

Materials

Cork from a wine bottle
Double-sided sticky tape
Pieces of broken porcelain
Fine concrete
Sand

Tools

Scissors
Mixing pail
Trowel
Old tablespoon
Wooden or foam spacers (if needed)
Plastic sheet
Hammer
Coarse sandpaper

Instructions

1. Cut the wine cork down to about 2 cm (³/₄ in) and attach it to the bottom of the larger mould using double-sided tape – this will create a draining hole in the bottom of the concrete pot. Place the porcelain pieces onto the tape with the design facing downwards and then cut around them so the tape covers the whole of their surface but doesn't stick out beyond it . Attach the porcelain to the inside of the larger mould.

2. Mix the dry concrete with a little water until it forms a spongy, viscous paste. Pour a little concrete into the large container to reach up to the top of the cork. Shake the mould lightly to even out the concrete.

3. Fill the inner mould with sand and place it on top of the cork in the middle of the large container. Spoon the concrete into the space between the moulds, leaving a little space at the top.

4. Hold both moulds and shake them lightly so that the concrete runs down and the air bubbles come up to the surface. Make sure that the inner mould doesn't slip to the side or the pot will have walls of varying thicknesses (add wooden or foam spacers between the two moulds in order to keep the inner mould in place, if liked). Cover with a plastic sheet and leave to harden for two days.

5. Empty out the sand and remove the inner mould. Stretch the outer mould so that the concrete loosens from the plastic and then press out the concrete pot from underneath. If the mould is stuck, you can try pouring a little hot water onto the edge between the plastic and the concrete. If the pot still won't release you may have to cut into the plastic bowl to remove it.

6. Remove any tape still stuck to the porcelain and scrape away any concrete that's run over it. Carefully tap the cork out using a hammer and even out the top edge of the pot with sandpaper.

Porcelain tips

- For a rectangular pot, it is best to use flat porcelain pieces. Avoid curved or convex ones.
- Use mosaic cutters, available from craft stores, to cut porcelain pieces into the desired shapes.
- Wrap porcelain in a rag or newspaper before breaking into pieces to avoid scattering shards around your work area.

- Use wide double-sided tape, sold in paint and building materials stores, to keep the porcelain pieces in place. The tape will often still be stuck to the porcelain when you remove the mould, but can easily be pulled off. This sort of tape can be quite sticky so your scissors will get sticky too, but you can wash them with methylated spirit.

ABOVE: Brazier by Lottie Svedenstedt, cast in a round plastic container. Porcelain pieces were placed close together on the edge and pressed into the wet concrete.

BELOW AND RIGHT: Porcelain pieces from the 1940s adorn the underside of these 'nostalgia pots' by Lottie Svedenstedt. The pieces were attached to the finished concrete bowl using tile adhesive.

ABOVE: Porcelain pieces from Bers and Prunus tableware sets by Stig Lindberg decorate a little candleholder by Annica Leijon-Ri⟩ The candleholder was cast in a plastic dessert container.

Flowerbed face

There are moulds for concrete everywhere – you just have to look for them! I found a smooth plastic face mask in a craft shop, which was meant to be painted and decorated for a masquerade or for Halloween. The mask was excellent to use as concrete mould, with a little support from wet sand. With a steel rod cast into it, the result was a fun and surprising decoration for the flowerbed.

Mould

Face mask made of plastic, rubber or
   thick card
Wet sand

Materials

Strong tape
Steel rod measuring about 1.3 m (1.3 yd)
Fine concrete or another
   type (I used glass
   fibre concrete, which
   creates a light surface)

Tools

Mixing pail
Trowel
Plastic sheet
Coarse sandpaper

Instructions

1. Tape over any holes (eyes, mouth etc.) in the mask from the outside so concrete cannot escape from them (on my mask I also had to tape the top of the forehead to prevent the concrete from running out).

2. Place the mask in a valley of wet sand to support the sides and underneath. Make a hole in the chin and push the steel rod through the hole into the middle of the mask. Wedge more sand under the other end of the rod so that it lies horizontally.

3. Mix the dry concrete with a little water until it forms a spongy, viscous paste. Pour it into the mask. Shake the mould a little so the concrete evens out and runs in under the end of the rod. Fill the mask with as much concrete as possible.

4. Shake the mould lightly to bring all the air bubbles up to the surface. Cover with a plastic sheet and leave to harden for two days.

5. Remove the mask and file the edges of the face with coarse sandpaper.

A hypertufa pot positioned in the shade or half-shade will soon be covered in beautiful lush moss.

Hypertufa

If you mix peat into concrete, you get a more porous material called hypertufa. It takes longer to dry than regular concrete, so while still half-set you can shape it in a completely different way than you can regular concrete.

It is best to mix peat with sand and cement rather than fine concrete. Cement and sand are available at building materials stores and non-fertilized peat is available in sacks from garden centres and plant stores.

I made the round urn to the right in a mould made from a cardboard tube with a tall plastic pot as the inner mould. When the hypertufa had hardened somewhat but was still soft I pulled away the outer cardboard tube and sculpted the ridges with a loop (a modelling tool used by potters).

You can make square troughs in the same way using cardboard boxes of various sizes. The box softens a little from the wet concrete and gives the finished trough a rounded form. Wrap a few layers of packing tape around the outer box so it doesn't lose its shape completely.

Mould

A piece of cardboard tube about 25 cm (10 in) in diameter for the outer mould and a tall plastic pot about 16 cm (6¼ in) in diameter for the inner mould.

Materials

Standard cement
Sand
Non-fertilized peat
Cork from a wine bottle
Weight, such as a house brick
Piece of wood (if needed)

Tools

Plastic bag
Loop or modelling knife
Mixing pail
Trowel
Old tablespoon
Spray bottle
Plastic sheet
Corkscrew

Instructions

1. Place a plastic bag on a flat surface and put the cardboard tube on top of it. Thoroughly mix 1 part cement with 1 part sand and 1.5 parts peat, tearing apart any large pieces of peat. Add a little water at a time and blend until the mixture is is thick and spongy.

2. Pour the mixture into the tube to make a bottom layer 3–4 cm (1¼–1⅗ in) thick. Press the cork into the bottom layer to create a draining hole in the finished pot. Fill the plastic pot with sand and place it on top of the cork and bottom layer.

3. Layer the hypertufa into the space between the cardboard tube and plastic pot. Pack it in so that the mixture goes down all the way. Fill it almost all the way up to the top and place a weight on top (with a piece of wood underneath, if needed). Cover with the plastic sheet and leave to harden until the following day.

Newly made round hypertufa urn. In the background is an older hypertufa trough that has dried and lightened.

4. Moisten the cardboard tube using the spray bottle then remove the tube. Make ridges or another pattern in the hypertufa using the loop and spray with water to prevent it from drying out. Cover with the plastic sheet and leave to harden for three days.

5. Carefully turn the urn over and empty out the sand. Pull out the inner mould and remove the cork using the corkscrew. Cover the urn with the plastic sheet and leave to harden for a few more days.

Rustic garden

If you dream about having a beautiful, rustic garden then hypertufa is the right material for you. With the help of hypertufa troughs, urns and benches you can quickly create your own bewitching garden in a shady corner of your plot. The peat makes hypertufa's surface very porous and gives the material an aged look almost immediately. And as hypertufa takes a few days to become really hard, you'll have time to round out the edges on your finished designs while they are still soft.

The name 'hypertufa' comes from 'tufa', which is the soft, volcanic rock this material resembles. Moss grows well on hypertufa objects in shady locations and if you want to hurry its growth, brush the surface with soil mixed with a little yogurt.

TOP RIGHT: This bench was made using a horizontal piece of leca and two parallel leca blocks, which were then covered in hypertufa. The round urn was cast in a tall plastic pail with a paper tube as the inner mould.

NEAR RIGHT: This armchair has a framework of mortared leca blocks, covered in hypertufa.

MIDDLE RIGHT: Hypertufa troughs cast in taped cardboard boxes of varying sizes.

FAR RIGHT: An aged-looking hypertufa trough. The corner was shaped before the hypertufa dried.

These small decorative mushrooms for the garden are cast in sand and the stems are made using cardboard tubes. If the mushrooms are kept in the shade, they will eventually become overgrown with moss. You can help the moss grow by brushing the surface with yogurt mixed with soil. Cast pieces of steel rod in the mushrooms so they will stand steady in the grass.

Mould
Wet sand in a tub
3 cardboard tubes (such as toilet roll inners)

Materials
Cement
Sand
Non-fertilized peat
3 lengths of steel rod

Tools
Mixing pail
Trowel

Instructions

1. Make 3 indentations in the sand to form the cup of the mushrooms.

2. Thoroughly mix 1 part cement with 1 part sand and 1.5 parts peat, tearing apart any large pieces of peat. Add a little water at a time and blend until the mixture is thick and spongy. Fill each indentation with the paste and press the pieces of steel rod into them.

3. Thread a cardboard tube over each length of steel rod and press the tubes down into the hypertufa. Fill the tubes with hypertufa and leave to harden for two days.

4. Pull out the mushrooms and brush or rinse away the sand. The cardboard will have softened from the moisture and can just be pulled away.

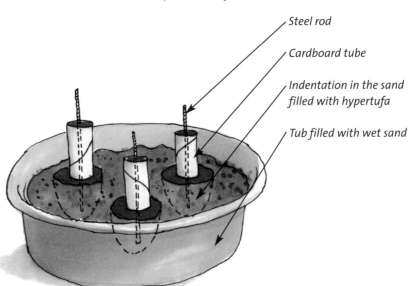

Steel rod

Cardboard tube

Indentation in the sand filled with hypertufa

Tub filled with wet sand

Bench and birdbath cast on a mound of sand by Annika Agdell

Benches

Until a few years ago, I dragged a few wooden benches and stools out into the garden each summer to sit on. Often, I had to scrape and repaint them, and then it was time to put them back in the garage for the winter. Nowadays, I cast benches in concrete instead: they stand outside all year round and need no maintenance.

I used 12 mm (½ in) thick form plywood to make the mould for the simple, austere bench to the left, and cast it in one piece. The mould was screwed together well and placed on its side before the fine concrete was poured in. The concrete surface turned out a little porous since it was hard to shake the mould to get rid of the air bubbles.

I used an old drawer to make the three parts of the little concrete bench on the right. First I made the top, then I placed a separator in the drawer and cast the two pieces for legs, and finally I glued the pieces together with a strong glue for outdoor use.

Mould
Pieces of formwork plywood

Material
Two pieces of thin steel rod, about 8 mm ($5/16$ in) in diameter

Fine concrete (for the bench in the picture, which is 42 cm/16.5 in, tall, I used around 1.5 sacks weighing 25 kg/55 lb each)

Tools
Saw

Screws and screwdriver

Plastic sheet

Bolt cutter or hacksaw

Mixing pail

Trowel

Piece of wood

Weights, such as bricks

Sponge

Rasp or coarse sandpaper

Instructions
1. Think this project through thoroughly before you begin work and write down the measurements of the bench: how tall, wide and long you want it. Make sketches on the pieces of plywood that will be sawed, or use the measurements offered here. Remember that the pieces will overlap each other when they are screwed together, as in the picture. Measure and saw the pieces.

2. Screw the pieces together as in the picture, so the mould is as strong and tight in the joints as possible. Place the mould on its side on top of the plastic sheet, on a smooth, level surface.

3. Bend two thin pieces of steel rod so you can place them right in the middle of the mould. You can bend the rods yourself or you can cut them to the right length with bolt cutters or saw them with a hacksaw.

4. Pour a few litres/quarts of concrete powder into the pail and add water a little at a time. The paste should be spongy and viscous. Pour it into the mould and press the concrete down with a piece of wood. Make another batch when necessary – you will need several batches even for a small bench. Place the two pieces of steel rod directly into the concrete.

5. Place the weights on the mould so it stands steady and doesn't gape open. Fill it all the way up and smooth out the surface with a moist sponge. Cover with the plastic sheet and leave to harden for three days.

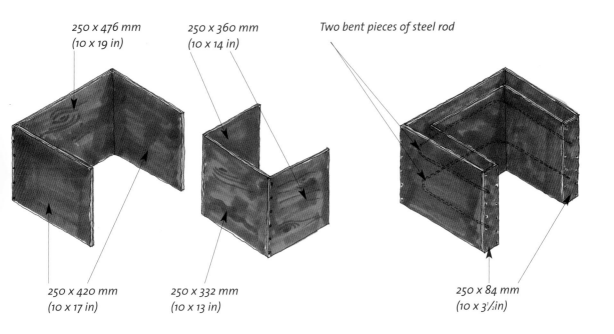

250 x 476 mm
(10 x 19 in)

250 x 360 mm
(10 x 14 in)

Two bent pieces of steel rod

250 x 420 mm
(10 x 17 in)

250 x 332 mm
(10 x 13 in)

250 x 84 mm
(10 x 3½ in)

6. Unscrew the wooden mould and remove. Stand the bench upright and carefully knock away any concrete that has run out from the joints. File any rough edges with a rasp or coarse sandpaper.

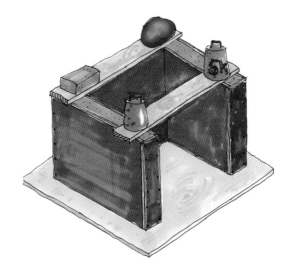

Mould

Drawer or other wooden frame
A separator to create two compartments
in the drawers

Materials

Wood lacquer (if needed)
Film canisters or other round plastic
tubes to create holes (if desired)
Wide double-sided tape
Fine concrete
Strong glue for outdoor use

Tools

Weights, such as bricks
Marker pen
Mixing pail
Trowel
Plastic sheet
Coarse sandpaper
Hammer (if needed)
Screws and screw-driver

Instructions

1. If using untreated wood, lacquer the drawer
or wood pieces before use. If the drawer is
bottomless, use any smooth, lacquered wood
or laminate to create a base, or place directly
on a plastic-covered table. Put weights
around the frame so it doesn't gape open.

2. To make holes in the stool (as in the picture)
tape film canisters to the bottom of the
drawer using double-sided tape (the tape
must not stick out beyond the canister or it
will leave an impression in the concrete).
Make a mark about 3 cm (1 in) from the base

of the mould – this will be the thickness of
the bench's seat.

3. Pour a few litres/quarts of concrete powder
into the pail and add water until the paste is
thick, spongy and viscous. Pour it into the
mould up to the mark. Make another batch if
necessary. Cover with a plastic sheet and leave
to harden for two days.

4. Turn the mould over and shake out
the concrete. Smooth the
edges and corners with
coarse sandpaper.
Tap out the film
canisters using a hammer.
Scrape away any remnants of the concrete
from the mould.

5. Screw down the separator so that you get
two compartments in the drawer/frame.
Then cast the legs in the same way.

6. Glue the legs and top together with a strong
glue for outdoor use.

Mini-table

Using the same old drawer I also cast three pieces for a little coffe table that I could place beside my chair. I glued the pieces together with a strong glue for outdoor use.

*Deckhopper chair
by Mathias Claerhout.*

Candlesticks

You can find many exciting packages to use as concrete moulds in the grocery store. I made candlesticks of different heights from round potato snack tubes. The decoration is made from circular pieces cut out of a plastic lace cloth.

Mould
Cardboard tubes

Materials
Plastic crocheted cloth
Spray glue
Fine concrete
Tea lights

Tools
Saw or bread knife
Kitchen paper
Scissors
Mixing pail
Trowel
Old tablespoon
Stones, to use as weights
Coarse sandpaper

Instructions
1. Saw the paper tubes to the required height and clean the insides with kitchen paper.

2. Cut out rounds from the plastic lace cloth and glue them inside the tubes using spray glue. Place the tubes on an even surface.

3. Mix the concrete powder with water, a little at a time, until it forms a viscous paste. Spoon the concrete down into the tubes, filling almost up to the top. Shake the tubes a little so the concrete sinks down and the surface evens out.

4. Carefully press the the tea lights down into the concrete. Place a weight, such as a small stone, on top so the tea lights don't rise up. Leave to harden for two days.

5. Remove the paper tubes and cloth pieces. If there are any bits of paper left on the concrete, you can remove them by placing the whole candlestick in a pail of water. If necessary, even out the upper edge of the candlestick with coarse sandpaper.

The lacy pattern became more visible when I glazed the candleholders white. For more about white-glazing concrete, see page 55.

Variations on a square

I'm not so good at carpentry, but I did manage to make a little wooden square without too much trouble. I have used it to cast many small tiles, some with depressions for candles or soap, but most with different decorative details cast into them, such as pieces of tile, wood or mirrors. 'Lovely, Malena,' my friends say, 'but what do you need all these pot holders for?'. But it isn't about me needing pot holders. It's about the desire to create. About an idea that has be tried. About enthusiasm and the joy of creation.

If you like carpentry, you can make a larger mould with more compartments, so you can cast lots of tiles at the same time. There are many variations on a square. You can experiment with casting using corks, shards of china and coloured glass. With a small mould, it is extra important to be careful with measuring, sawing and screwing. The slightest flaw will leave an impression in the finished concrete product.

Waxed paper letters from the craft shop made the imprint in this concrete tile by Kristina Digman.

Mould

Pieces of formwork plywood

Materials

Tile
Double-sided tape
Fine or another concrete
Small, adhesive furniture pads

Tools

Saw
Screws and screwdriver
Spirit level (if needed)
Mixing pail
Trowel
Marker pen
Plastic sheet
Thin steel spatula (if needed)
Coarse sandpaper

Instructions

1. Saw the pieces of plywood and screw them together to create a mould that is 13 x 13 cm (5 x 5 in) square and about 4 cm (1½ in) tall. The mould should be without gaps so the cement won't run out through the joints. Use a spirit level, if needed, to ensure that the mould is on a flat surface.

2. Place double-sided tape over the entire good side of the tile and attach it to the bottom of the mould. The tape cannot stick out beyond the tile or it will leave an impression in the concrete. Make a mark to show how thick you want your slab; here it's 2.5 cm (1 in).

3. Mix the concrete powder with water into a viscous paste and pour it in up to the mark. Cover with a plastic sheet and leave to harden for two days, then unscrew the mould. If the concrete slab is stuck to the bottom of the mould slip in a thin steel spatula to loosen it. You can file away rough edges and corners with coarse sandpaper.

4. Place small, adhesive furniture pads on the slab so it doesn't scratch the table and don't forget to clean the square mould very well before you cast the next tile.

A wooden coaster was taped to the bottom of the square mould to create this effect.

This pattern was created by attaching 16 small square wooden plugs to the bottom of the mould using double-sided tape. Don't forget to hit out the plugs from the front in order to avoid chipping the edges around the holes.

The holes in this design were created using four round film canisters.

To make this mirror tile I taped a piece of cardboard to the bottom of the mould before adding the mirror in order to raise the mirror away from the surface of the concrete. However the cardboard piece absorbed water and made the edges next to the mirror sandy. Next time I will oil or cover the cardboard in plastic first and place a wire hook in the concrete so I can hang the mirror when it's finished.

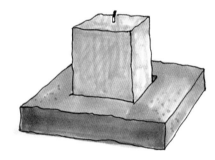

Here I positioned a plastic container inside the mould to create the depression in the concrete tile to hold a square candle.

Concrete soap dish, made in the same way as the candle holder above.

I made the stand for this tile using a smaller mould, then cast the tile itself in a regular mould before glueing the pieces together.

Resistant placemats

Just for fun I made two rain-resistant placemats for the table in the garden. On one I placed a knife and fork and some blue and white porcelain, and the other got an Asian touch with chopsticks and mosaic pieces. I cast the placemats upside-down in a little wooden case, using the same principle as for the squares on page 90.

Mould

Small plastic or wooden case (if the wood has not been treated, it should be lacquered first)

Materials

Porcelain pieces, silverware, chopsticks or other decorations
Wide double-sided tape
Fine or fibre concrete

Tools

Hammer (if needed)
Marker pen
Mixing pail
Trowel
Plastic sheet
Thin metal spatula
Coarse sandpaper

Instructions

1. Attach the decorations to the bottom of the case using double-sided tape. Mosaic and porcelain pieces should be as flat as possible and forks need to be flattened using a hammer. The tape should cover the entire decoration but not stick out, otherwise it will leave a tape impression in the concrete.

2. Make a mark about 1.5 cm (1 in) up from the base of the case – this will be the thickness of the placemat.

3. Sprinkle several litres/quarts of concrete powder into the pail and add a little water. Mix until it is an even, viscous, spongy paste.

4. Pour the concrete into the mould up to the mark and shake so the concrete lies evenly and the air bubbles rise to the surface. Cover with the plastic sheet and leave to harden for two days.

5. Stretch the mould a little, turn upside-down and shake out the tile. If it doesn't come out, you can remove one side of the mould and slip a thin metal spatula between the concrete and the bottom of the wood.

6. Remove any tape that remains and scrape away any concrete that's run over the decoration. If necessary, file the edges with coarse sandpaper.

Garden tiles

Concrete is a resistant and durable material, perfect for use on paths and outdoor areas. For small spaces you can cast unique tiles and experiment with moulds and patterns using leaves or mosaics.

The simplest way to make moulds for tiles is using plastic containers, then it is really easy to cast tiles – just place all the containers on a level surface, mix a good portion of concrete to pour into them and wait two days. However try to avoid using biscuit containers as they can create rounded corners on the finished tile. Also bear in mind that reliefs and punched company designs or trademarks on the bottom of the container will leave impressions in the concrete.

If you want to make tiles with tighter designs, sharper corners and perhaps in larger sizes, you can make frames out of wood. Smooth tiles can be made by casting upside-down against a smooth and level bottom layer in the mould.

Moulds made of plastic and wood, ready for casting. The plastic moulds contain leaves that will create an impression in the concrete. Photographed by Ingegärd Bodén.

I cast tiles for the pathway through my greenhouse in a small wooden frame with a formwork plywood base. I bought the gingerbread moulds at a flea market.

Mould

Leftover pieces of formwork plywood
Gingerbread or other biscuit cutters
Weight, such as a stone or brick (needs to be larger than the biscuit cutter)

Materials

Transparent silicone (if needed)
Fine concrete
Oxide red pigment

Tools

Saw
Screws and screw-driver
Mixing pail
Trowel
Plastic sheet
Thin steel spatula or knife (if needed)
File

Instructions

1. Saw the pieces of plywood to the required size. Do this carefully – the more precise and tight the mould is, the better the corners and edges of the tile will look. Screw the mould together well so there are no gaps. (If gaps do appear, cover them with transparent silicone.)

2. Place a biscuit cutter in the middle of the wooden mould. Mix a little concrete and add a drop of pigment so the paste turns red. Fill the biscuit cutter with the red concrete and place a weight on top. Wipe-up any red-coloured water that leaks out from under the biscuit cutter. Leave to harden for a few hours.

3. Mix a few litres/quarts of dry concrete with a little water until the mixture forms a firm consistency that can be shaped into a ball. Pour the concrete into the wooden mould so the biscuit cutter is buried. As a guide, a tile that is 30 x 30 cm (12 x 12 in) ought to be at least 4 cm (1½ in) thick. Press down using a trowel so the concrete gets into the corners. Shake the mould lightly so that the air bubbles come to the surface. Cover with a plastic sheet and leave to harden for two days.

4. Unscrew the mould and pull out the tile. If it is stuck, slip in a thin steel spatula or knife and loosen it. File away any concrete that has slipped out from between the joints. The biscuit cutter should not be visible and can remain in the tile.

I wanted to make a floor for my greenhouse using 12 concrete tiles, 4 cm (1½ in) thick and 30 x 30 cm (12 x 12 in) square. I screwed together a lattice framework with six compartments so I could make six tiles at a time. I placed the wooden lattice directly on top of a table covered with an opened plastic rubbish bag. This worked well but I had to stretch the bag out and tape it very tightly as the slightest wrinkle can leave an impression in the concrete.

Mould
Remnants of wood

Materials
Wood lacquer (if needed)
Peony leaves
Spray glue
Fine concrete

Tools
Saw
Screws and screw-driver
Plastic rubbish bag
Set square
Weights, such as bricks (if needed)
Marker pen
Mixing pail
Trowel
Plastic sheet

Instructions

1. Measure and saw the wood for the mould to the required size. Any untreated wood that seems absorbent should be lacquered before use. Screw the wood together well so it is as tight as possible, with at least two screws in each part (use a set square to ensure 90-degree corners). The pieces of wood in the middle of the mould can usually be pressed down so they stay in place without screws.

2. Rip open a rubbish bag and tape it so it is stretched out across a table surface. Place the mould on top. You may need weights in the corners, such as bricks, so the mould stays steady and doesn't gape open towards the surface underneath.

3. Spray-glue the peony leaves with their backs upwards (the side where the veins show most clearly) directly onto the plastic. I placed them so they formed a creeper that stretched from one tile to the next. Make a mark about 4 cm (1½ in) up from the base on the inside of the mould, so you know how much concrete to pour in.

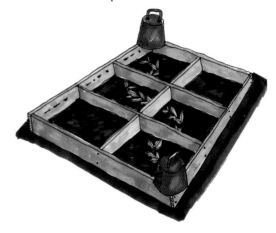

4. Pour a few litres/quarts of dry concrete powder into the pail. Add a little water at a time and mix until you have a very viscous paste – the consistency should be firm enough to form the concrete into a ball. Pour the concrete into the compartments in the mould, up to the mark. Pack the concrete in well using the trowel so there are no air pockets. Mix more concrete as needed. Cover with a plastic sheet and leave to harden for two days. Lift the plastic and pour water on the concrete a few times during the second day.

5. Remove the plastic and unscrew the mould. Carefully remove the peony leaves. Scrape the mould clean and change the plastic on the table before the next casting.

I used different types of concrete for these tiles, which is why they are of varying shades. Use frost-resistant concrete for outdoor projects.

This irregular tile was cast upside-down directly onto a slice of masonite. You can also cast it on top of a piece of lacquered wood or on a durable, plastic-covered table. Remember that the surface of the tile will only be as smooth or rough as the surface used. The edges of the mould were made from flexible rubber supported by sand. You can also use a strip of veneer, a thin tray, flexible plastic or anything else that can be shaped into a puzzle piece and can stand upright when supported by the sand.

Mould

Wood, masonite or durable table
Wide strip of rubber, veneer or similar
Wet sand

Materials

Wood lacquer (if needed)
Fine concrete

Tools

Mixing pail
Trowel
Plastic sheet
Thin steel spatula or knife
Coarse sandpaper

Instructions

1. If you are using wood that seems absorbent, it should be lacquered first with wood lacquer. Mould the strip of rubber into a puzzle piece-shape on top of the wood, masonite or table and support it by packing wet sand around the outside.

2. Mix a few litres/quarts of dry concrete powder with a little water at a time until it forms a stiff paste. Pour the concrete into the mould – as a guide the tile should be at least 4 cm (1½ in) thick if it is 30 cm (12 in) in diameter. Cover with a plastic sheet and leave to harden for two days. Lift the plastic and pour water on the concrete a few times during the second day.

3. Remove the rubber strip and turn the tile over. If stuck, slip in a spatula or knife and loosen it. File any rough edges with coarse sandpaper. Bury the tile down to the level of the lawn, so you easily run the lawn mower over it.

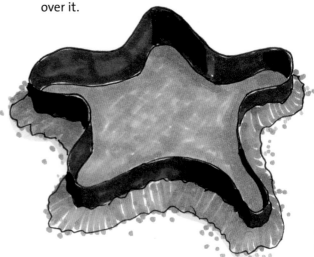

*Puzzle piece tile, just removed from
the mould, by Maria Jönsson.*

Concrete pillow

Not exactly something to place under your head when you're taking a nap in a hammock, this pillow looks attractive as it collects rain water in its indentation.

Mould
Strong plastic rubbish bag
Duct tape
Balloon or round boulder

Materials
Fine concrete

Tools
Mixing pail
Trowel

Instructions
1. Cut and tape the rubbish bag to make a 'pillowcase' in the desired shape, open on one side.

2. Blend a few litres/quarts of concrete powder in the pail and add water a little at a time. The mixture should form a spongy, viscous paste. Fill the bag with cement and tape it closed securely. Pull on the bag so the concrete squashes into the corners and is evenly spread.

3. Fill the balloon with water and place it or the boulder in the middle of the pillow to create the depression. Leave to harden for two days.

Concrete pillow by Ingegärd Bodén. Photographed by Ingegärd Bodén.

Coloured spheres

Candleholder by Inga-Lill Lager. The sphere was made using a toy ball.

George Little and David Lewis are two American artists who make concrete sculptures glazed in strong colours. In their famous garden outside Seattle large, colourful concrete spheres stand on pedestals, while others sit directly on the ground like forgotten dinosaur eggs. Enormous concrete gunnera and rhubarb leaves, painted lime green, orange and lemon yellow, are placed behind designed waterfalls where they shine along with the red canna leaves and emerald-green banana plants.

The dinosaur eggs really inspired me and I tried a number of different methods to create them – filling toy balls will cement or making moulds out of papier-mâché around balloons are just a couple of methods you can experiment with. The method I had the most success with was also the simplest: covering a giant balloon or beach ball with supple concrete putty. Not difficult at all!

Two concrete spheres cast around a giant balloon and a toy ball.

107

Mould

Giant balloon (or plastic toy ball)
Bowl or dish filled with wet sand

Materials

Concrete putty

Tools

Mixing pail
Trowel
Steel putty knife
Sponge
Large plastic bag
Coarse sandpaper
Water-based white primer or pigment
(and other colours, if desired)
Paintbrush (if needed)

Instructions

1. Make an indentation in the wet sand in the bowl. Blow the balloon up to 26 cm (10 in) in diameter (not larger, or it will burst) and place it in the sand with the knot facing down. Press the sand around it so the balloon is supported.

2. Mix a couple of litres/quarts of dry concrete powder with a little water at a time in the pail until it forms a supple, gluey mass.

3. Hold the balloon steady and place a 'hood' of concrete on top of it about 1 cm (³/₈ in) thick (if the concrete paste is the correct consistency it will stay on the balloon). Rotate the bowl so the balloon is evenly covered all the way around, otherwise it will tip over from the concrete's weight. Use a steel putty knife to even out the surface as much as possible (you have about 20 minutes before the paste begins to harden). Cover with the plastic bag and leave to harden overnight.

4. Turn the balloon over, mix more concrete putty and cover the rest of the balloon's surface, including the joint between the old and the new putty. Leave an opening around the knot of the balloon. Use the sponge to smooth out the surface. Cover the whole sphere with the plastic bag and leave to harden until the following day.

5. Poke a hole in the balloon to deflate it and sand the sphere with coarse sandpaper. I painted the inside of my spheres with diluted white primer mixed with a little marine blue and gold pigment. For more on painting concrete, see page 116.

I made this large sphere with an uneven opening around a beach ball, so that it resembled an eggshell. I strengthened the concrete with strips of fibreglass, which is available by the roll from hardware stores. To avoid impressions from the beach ball's seams, I used a wash of a runny concrete mixture on the inside. I experimented with different glazes to get the shimmering blue colour resembling a bird's egg. Now the giant egg rests under a walnut tree and collects rain water and falling leaves. In the winter, I simply empty it and place it under a shelter.

Mould
Beach ball
Wet sand

Materials
Concrete putty
Strips of fibreglass

Tools
Mixing pail
Trowel
Steel putty knife
Sponge
Large plastic bag
Coarse sandpaper

Instructions
1. Place the beach ball on a mound of sand with its valve facing downwards. Bank sand around it so that it stands steady.

2. Mix the concrete putty with water to form a pliable mass and shape a 1 cm (³/₈ in) thick 'hood' on top of the ball. If the concrete is really pliable, you can cover quite far up the ball's sides without the paste falling off. It is important to apply the concrete evenly to all sides, otherwise the ball will fall over. Cover with strips of fibreglass. Add another layer of concrete and even it out with the putty knife and sponge. Cover with a plastic bag and leave to harden until the following day.

3. Turn the ball over carefully and continue to cover with concrete and fibreglass. Smooth out the joint between the old and the new layers. Leave to harden until the following day.

4. Open the valve, let the air out and remove the ball. Mix a really loose batch of concrete and cover the inside of the sphere to remove the imprints from the ball's seams. Cover the whole sphere with a large plastic bag and leave to harden for a few more days before smoothing the surface with sandpaper.

5. To get the shimmering blue colour, paint with a thin glaze made from marine-blue pigment. (For more on painting, see page

After I made the beach ball-sized sphere, I wanted to experiment with even larger moulds. A work-out ball, such as a pilates ball, was a good mould. I anchored the ball in damp sand and formed a concrete 'hood' on it.

My neighbours thought I had seriously misunderstood what a pilates ball was for, but now they admire the giant eggshell shard, which I planted with houseleeks. I then washed and re-used the pilates ball.

Mould
Pilates ball
Wet sand

Materials
Concrete putty
Cork from a wine bottle
White water-based primer paint

Tools
Mixing pail
Trowel
Sponge
Large plastic bag
Sandpaper
Paintbrush

Instructions

1. Place the ball on a mound of sand, so it is as steady as possible.

2. Mix the concrete putty with water in the pail until it forms a supple mass. Shape a 2 cm (¾ in) thick concrete 'hood' on top of the ball (I made the edges pointy to resemble a broken egg shell). Press a cork into the middle to create a draining hole. Smooth out the surface with a well wrung-out sponge. Cover with the plastic bag and leave to harden for 2 days.

3. Loosen the concrete from the ball and remove the shard. Sand the outside with sandpaper and press out the cork. Paint the eggshell-shaped shard with diluted white water-based paint.

Notes
- Concrete spheres and shards cannot withstand water freezing in them and so should be kept dry and sheltered during the winter months.

- Check with your building supply store that the concrete you use will create a really leathery, plastic mass when it is mixed with water. Also check how frost-resistant it is.

- Do not begin with giant spheres. Start with football-sized ones until you get the hang of the technique.

- Spheres and shards are quite fragile, even though they are made of concrete.

The eggshell shard works well as a planter. The pedestal is made from a polished piece of leca.

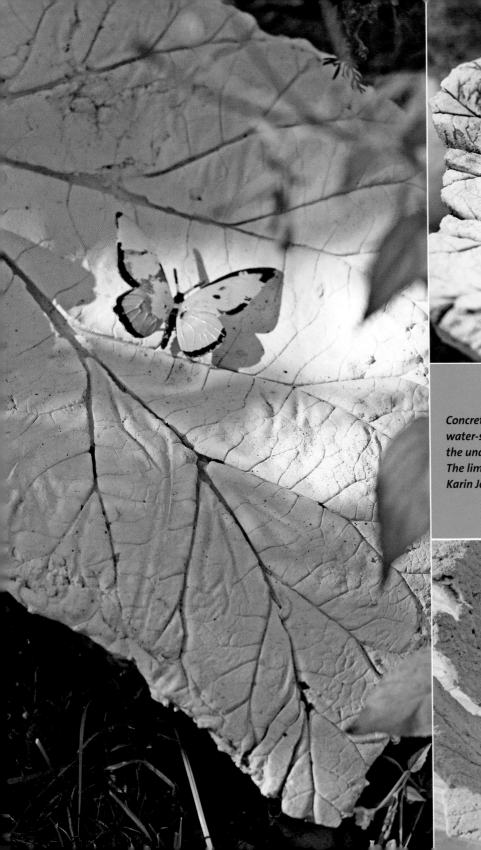

Concrete rhubarb leaves painted with water-soluble pigment. On the blue leaf, the underlying red colour shines through. The lime-green rhubarb dish was made by Karin Jonsson.

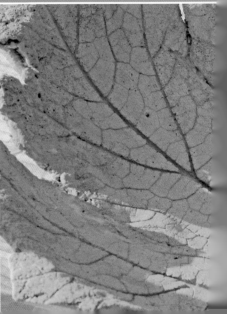

Colourful leaves

Many people have made concrete rhubarb leaf plates – a simple project for beginners that always turns out well. If you get tired of the grey surface, leaf plates are excellent for experiments with painting.

The artists George Little and David Lewis use water-based colours for their brilliant concrete leaves in bright colours. The water-based pigment is sucked into the concrete and creates a matte glazing effect. You can try painting layer on top of layer in different colours in order to get shimmer and depth.

In their book *A Garden Gallery* (Timber Press, 2005), the two artists tell how it took many hours in their studio to get just the right turquoise blue for a certain sculpture. Don't give up if you don't succeed the first time. It can take time to find the right colour, you just have to experiment.

Liquid pigment, available in bottles from hardware stores, is long-lasting and is easy to mix into water-based primer. Dry, powdered pigment (available from art supply stores) also works well. Just be thorough when mixing the powder with the primer. I usually experiment with both white primer and colourless glaze, which lets underlying colours shine through.

Any leftover paint can be kept in glass jars with lids or in tins covered with plastic. Save ice cream and lollipop sticks to use for mixing or buy wooden spatulas at hobby and craft stores.

Materials
White, water-based primer
Tins, glass jars or paper cups
Pigments in various colours, in powdered or liquid form
Concrete leaf plates
Colourless, water-based glaze (if liked)

Tools
Wooden spatulas or ice cream sticks
Paintbrushes

Instructions
1. Pour about 100 ml (3½ fl oz) white primer into a jar and dilute with water until it resembles thin cream. (When the paint is diluted enough, it will penetrate into the concrete more easily and won't just lie on the surface.) Separate the mixture into three containers. Mix a different colour into each container, such as yellow, blue and red.

2. Paint the whole leaf with one colour and allow to dry, then paint it with another colour and immediately dry the leaf off. Repeat with the third colour. The first colour will be the main one while the second and third colours will show in any indentations.

3. Colourless, water-based glaze looks milky in the jar but is completely transparent when it dries. Try mixing it with different colours and paint one layer on top of another (letting it dry in between), so the colours shine through one another.

CONCRETE LEAF

Rhubarb leaves, with their prominent veins, work well for casting. Also try casting leaves from hollyhock, gunnera, hosta and lady's mantle plants. Make several at a time so that you can experiment with painting.

Mould
Sand
Rhubarb or other leaves

Materials
Plastic sheet
Fine concrete

Tools
Mixing pail
Trowel

Instructions
1. Working on the ground, on top of a piece of plastic, make a level mound of sand about as large as the rhubarb leaf.

2. Place the rhubarb leaf on the sand with the back facing upwards. Slice off the stalk so that it doesn't stick up and make a hole in the plate.

3. Pour a few litres/quarts of dry concrete into the pail. Add a little water and blend until it becomes thick and malleable.

4. Place a 2 cm (¾ in) thick layer of concrete on the leaf. Don't let the concrete go outside the leaf's edge as concrete that is cast against sand becomes granular. Level the top so the finished rhubarb plate will stand steady. Cover with the plastic sheet and leave to harden for 2 days. Spray the concrete with water a few times during the second day.

5. Remove the plastic, turn the plate over and remove the leaf. Leave the plate to dry for a few more hours before painting.

Garden gnome

Not long after Christmas, I bought a discounted plastic lamp in the shape of a gnome. It was easy to remove the lightbulb-holder and then I was left with a hollow gnome made of thin plastic, open at the bottom. It was a perfect casting mould, I thought, and I poured in concrete. But when the concrete had hardened, the mould was stuck and I had to break it off, piece by piece. Nevertheless, I think it was well worth the trouble to get my own concrete garden gnome, which can be left outside no matter what the weather, year after year.

Keep your eyes out for hollow plastic moulds to cast in, such as dolls, bath toys and figurine-shaped sweet packages.

Mould
Hollow gnome or other hard plastic figurine
Pail with sand

Materials
Strong tape (if needed)
Fine concrete
Plastic bag

Tools
Mixing pail
Trowel
Clippers or strong scissors
Chisel
Sandpaper

Instructions

1. Ensure the mould is open at the bottom so you can get the concrete in. If not, you can cut a hole with clippers or strong scissors. If there are any other holes in the mould, it is best to tape them shut from the outside using strong tape, so concrete can't leak out.

2. To get my gnome-mould to stand steady with the bottom up, I placed it into a pail filled with sand. Mix concrete powder with a little water into a spongy, viscous paste. Pour into the mould. Shake the mould so the paste runs down into all the nooks and the air bubbles come up to the surface. Cover with a plastic bag and leave to harden for two days.

3. Cut out a hole in the mould using clippers or strong scissors. Break the mould away in pieces, using the chisel. Work from the back as it is easy to make scratches in the concrete when doing this. File any scratches with sandpaper.

The plastic lamp I used to cast my garden gnome.

Flowerpot with relief

I have made many experiments with leaves, feathers and old lace tablecloths in my moulds, in order to make designs on concrete surfaces. Often the impressions have been both beautiful and detailed, but how could I achieve a raised relief? I bought a piece of linoleum from a craft store and, using a lino knife, I carved out a picture of stylized bamboo copied from a book of Japanese ink drawings.

I put the linoleum inside the mould for a tall, straight pot. On the first attempt the water in the concrete loosened the linoleum so there was barely any relief at all. For the second attempt, I oiled the linoleum piece and got a nice relief, but a very porous concrete surface, which is expected when you oil the mould.

For the third try, I lacquered the linoleum and finally got a really nice relief that looks particularly good when the light falls on it. I made the mould for a 25 cm (10 in) tall rectangular flower pot out of formwork plywood. For the inner mould, I used a tall coffee cup (the kind you get from take-away coffee bars). The result was a rectangular pot with a round planting hole.

Mould
Pieces of formwork plywood (regular plywood should be lacquered first)
Tall coffee cup, plastic pot or paper tube

Materials
Piece of linoleum with a carved design
Transparent lacquer (wood or craft)
Wood glue
Fine or other concrete
Cork from a wine bottle
Sand

Tools
Saw
Screws and screwdriver
Linoleum knife
Mixing pail
Trowel

Straps or rope
(if needed)
Plastic sheet
Hammer
Coarse sandpaper

Instructions

1. Lacquer the linoleum and allow to dry.

2. Measure and cut four pieces of plywood and a base for the pot's outer mould and screw together with at least four screws in every joint. Cut the linoleum so that it just fits against one of the outer mould's internal sides, with the back towards the mould. Glue in place with wood glue and allow to dry. Check that the inner mould fits in the outer one. It is important that the flowerpot's walls are at least 1 cm (⅜ in) at the thinnest part, otherwise the concrete can crack.

3. Mix a couple of litres/quarts of concrete powder with a little water into a spongy paste. Pour about 2 cm (¾ in) into the mould. Press in half the cork to create a draining hole in the bottom of the pot. Fill the inner mould with sand and set it on top of the cork and concrete. Fill the space between the inner and outer moulds with concrete, mixing more if the first batch isn't enough. If you are making a large pot, over 40 cm (16 in) tall, place straps or ropes around the mould to support it as wet concrete is very heavy and can crack the mould. Cover with a plastic sheet and leave to harden for two days.

4. Pour out the sand and pull out the cup or pot. Unscrew the mould, remove the pot and hit out the cork with a hammer. Smooth the edges with coarse sandpaper.

Relief experiment using rosehips, poppy seed cases and wheat. I pressed halves of each into air-drying craft clay using a rolling pin. When the clay had dried I removed the halved rosehips, poppy seed cases and wheat, and then lacquered the clay as it isn't water-resistant and would otherwise have dissolved in the wet concrete. I put the dry and lacquered clay tablet into a plastic container with the impression facing upwards and then poured concrete over. The result was a decorative tile with an outwards-facing relief.

Fruit and figurines

When you've cast tiles, pots and plates, and mastered the techniques, it can be interesting to try more complicated shapes such as fruit and figurines. You can find different types of rubber, silicone, plastic and plaster in craft stores which are made especially for creating concrete moulds.

Vinamold is a type of meltable vinyl plastic that works really well for making concrete moulds. It is rather expensive to buy but can be re-used, so when you get tired of a mould you can melt it down, make a new mould and start all over again.

Vinamold should be warmed in a pot on the stove until it has a syrup-like consistency. It can then be spooned or brushed onto the object you want to cast. I made the moulds for the concrete fruits pictured here by placing real apples and pears in a pail and pouring melted Vinamold around them.

Real and concrete fruits
photographed by Jakob Skote.

Mould
Vinamold
3–4 equal-sized fruits

Material
Fine concrete

Tools
Saucepan
Wooden spoon
Heat-resistant bowl
Mixing pail
Trowel
Coarse sandpaper

Instructions

1. Cut the Vinamold into pieces and melt in the saucepan over a low to medium heat. (You can also melt it in the oven in an ovenproof pan, but I think you have more control when it's on top of the stove.) Do not allow the Vinamold to boil and keep a window open for ventilation. Stir until the Vinamold is viscous and syrup-like and without lumps.

2. Remove the pan from the heat and allow the Vinamold cool slightly (it shouldn't be so hot that it begins to bubble when you pour it, but it should still be runny). Pour a thin layer, about 1 cm (3/8 in) thick, into the bowl then put the pan back on the stove so it doesn't harden (this can happen quickly).

3. When the bottom layer has dried place the fruit on top with space around each so they don't touch each other or the walls of the bowl. Pour Vinamold around the fruits until they are almost covered.

4. When the Vinamold has completely cooled turn the pan over and shake out the mould. Vinamold is elastic and you can usually press the fruit out of the mould (the fruit will usually go pulpy from the heat). Make sure the inside of the mould (where the fruits were) is dry and clean before you pour in the concrete.

5. Mix the concrete powder with water into a viscous paste and pour into the moulds. Shake the moulds so the concrete goes all the way down and the air bubbles escape. Leave to harden for two days.

6. Turn the mould over and press out the concrete fruits from underneath. Be extra careful with small stems since they can break off easily and you may have to cut a hole in the mould to release them. File away any edges and bumps in the fruits using coarse sandpaper.

Concrete pumpkin cast in a plaster mould, apples and pears made using Vinamold and a little shell cast in a soap mould.

Two-component mould silicone is made from two types of silicone clay that are mixed together and moulded around a desired shape, such as a shell or fruit. The silicone hardens in around 15 minutes and forms a hard, rubber-like mould you can cast in many times. Unlike Vinamold, the silicone mould cannot be melted down and re-used.

Mould silicone is available from art supply stores. As this material is rather expensive it is best used for small castings.

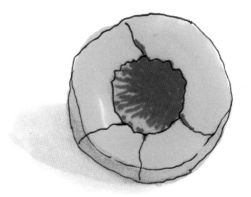

Real and concrete walnuts.

Mould rubber or mould latex is a liquid rubber that can be brushed in coats onto the object you want to cast. After about 10 coats, which must dry in between (it usually takes a couple of days to make a finished mould), you have a mould of the object. The slightly flimsy yellow rubber mould is fine to cast in, but needs to be set in a mound of sand in order to keep its shape when the heavy concrete is poured in.

Mould

Mould rubber or mould latex (available from craft stores)
Porcelain frog or other figurine
Bowl containing moist sand

Material

Piece of plastic or newspaper
Fine or another concrete

Tools

Paintbrush
Soap
Mixing pail
Trowel

Instructions

1. Place the frog on plastic or newspaper on a sturdy surface. Shake the bottle containing the mould rubber. Brush a thin layer over the entire frog, except for the bottom. Wash the brush in lukewarm water using soap.

2. Brush on the next layer when the first has yellowed and hardened. Continue in this way until you've coated about 10 layers. Don't forget to wash the brush every time.

3. Press the rubber mould (with the frog still in it) into a bowl of moist sand to make as good an impression as possible.

4. Carefully take the mould out of the sand. Remove the rubber mould from the frog and set the mould down in the sand again.

5. Mix the concrete powder with water to a spongy paste (or cast the rubber mould when you have leftover concrete). You don't need much concrete for a frog. Fill the mould (it doesn't need to be oiled) and shake the bowl a little so the concrete sinks down and air bubbles rise up. Leave to harden for two days.

6. Pull the rubber mould off the porcelain frog. You may have to cut the mould a little at the opening, but it can still be re-used.

Concrete melon cast in a plaster mould (see page 132).

Plaster is a cheap, easy-to-work-with material that is ideal for concrete moulds. It is best to make the mould in two halves that can be taken apart. If you are lucky, the plaster mould won't break and can be used several times.

Mould

About 5 kg (11 lb) ceramic clay
   (the cheapest kind)
Melon or other round fruit
About 5 kg (11 lb) modelling plaster

Materials

Cork from a wine bottle
Petroleum jelly
Rags
Fine concrete
Plastic bag

Tools

Mixing pail
Trowel
Flat putty knife
Old tablespoon
File

Instructions

1. Shape a layer of clay about 3 cm (1 in) thick around half of the melon. Make depressions around the top edge of the clay using the cork. Rub the top of the melon and the clay edge thoroughly with petroleum jelly.

2. Mix the plaster in the pail according to the directions on the package. Shape a 3 cm (1 in) thick layer over the exposed half of the melon. You have to work quickly as plaster begins to harden within 15 minutes.

3. Turn the melon upside-down and place it, with the plaster downwards, onto a layer of clay. Remove the clay from the melon. Wipe and remnants of clay away from the melon and plaster with a damp rag. Make a plug out of clay, about 5 cm (2 in) thick (or use a cork from a wine bottle), and place it on the edge of the plaster at one end of the melon – this will create the hole in the finished mould. Cover the exposed half of melon, the plaster edge and the plug with petroleum jelly. Mix more plaster and shape a 3 cm (1 in) thick layer over the exposed half of the melon. Allow to dry.

Rememberer to work quickly; plaster hardens in around 15 minutes.

4. Remove the plug. Insert a flat putty knife between the halves of the mould and prise them open carefully. When the mould halves are loose, you can grip between them and prise them apart.

5. Remove the melon and rub petroleum jelly on the inside and edges of the mould halves and the hole where the plug was. Put the mould together and wrap it tightly in rags. Place the mould with the hole upwards in a pail or tub and wedge rags under it to steady.

A cork from a wine bottle or a clay plug works well to create a hole in the mould.

6. Mix a loose concrete mixture and spoon it down into the mould through the hole using an old tablespoon. Shake the mould so the concrete runs down and any trapped air bubbles rise to the surface. Fill the mould, wait one minute and then check to make sure the concrete has sunk away from the hole. Add more concrete as needed. Cover with a plastic bag and leave to harden for two days.

7. Carefully prise the mould apart using a putty knife. Remove the concrete melon and file away any lumps or bumps while the concrete is still soft.

This mould was over-filled so it has a lump on one end that will have to be to be cut or filed away.

Materials

Craft clay
Petroleum jelly
Plaster, such as model plaster
Fine concrete
Cork from a wine bottle
Plastic sheet
Washing-up liquid

Tools

Mixing pails for plaster and concrete
Trowel
Thin knife or spatula
Small steel putty knife or chisel

The lion's face in white clay.

Plaster over half the lion's face.

Instructions

1. Make a lion's face out of craft clay. You could use pictures of lions as a reference or shape it from your imagination. Make a depression for the water to spurt out of the mouth. Allow to dry according to the package instructions.

2. Spread petroleum jelly on the clay face. Mix the plaster according to the instructions on the package. Place a 3-cm (1-in) thick layer of plaster onto the clay face. You have to work quickly as plaster hardens within 15 minutes. Allow to dry.

3. Loosen the plaster mould carefully using a thin knife or spatula. Position the cork at the lion's mouth to make the hole for the water to flow through. Rub the plaster mould with petroleum jelly.

4. Mix the concrete with water to form a stiff and malleable paste. Put it into the plaster mould. The cork should stick up out of the concrete and the edges should be at least 2 cm (¾ in) thick. Even the concrete out using the spatula. Cover with the plastic sheet and leave to harden for at least 2 days.

5. Slip the putty knife in between the plaster and concrete and try to loosen the plaster form. If it won't loosen, take the plaster off piece by piece using a putty knife or chisel. Wash away any plaster remnants with lukewarm water and washing-up liquid.

Mushrooms and birds are Annika Salminen's favourite motifs when making figures out of chicken wire and concrete. She shapes the chicken wire by experimenting and hangs it on a clothesline to see it from different angles. When she is happy with the mould, she fills it with newspaper and covers it with 1 cm (³⁄₈ in) of concrete. After the concrete has hardened, there is the time-consuming but fun job of positioning the mosaic tiles.

Fountain

Annika Salminen began making mosaic-covered concrete figures a number of years ago. Now she has developed her own method: she sculpts a mould out of chicken wire and newspaper, covers the mould with a thin layer of concrete and then places the mosaic tiles on top once the concrete is dry.

She has made many different shapes throughout the years: mushrooms, chickens, pillows and a large peacock. She uses frost-resistant glass mosaic tiles so the sculptures can be outside all year round. Annika thinks they are most beautiful when it rains and the tiles are wet, and that's how the idea for the mushroom-shaped fountain was born. The water comes out from the top, runs down the cap and drops into the bowl. In the bowl, there are three holes through which the water runs into a tub under the fountain.

About 1 cm (⅜ in) of concrete is placed on a mould made of chicken wire and newspaper. Press the concrete down in between the mesh of the chicken wire. Use several layers of concrete, so the chicken wire doesn't show.

From the side this fountain looks like a short mushroom in a bowl. Under the fountain there is a buried tub with water and below-ground a mini-pump. This pumps the water up through the pipe and out through the mouthpiece on top of the fountain. The water runs down the mushroom's cap, into the bowl and then through the three holes and back into the tub. The pump is connected to an earthed outlet and the cord can be hidden by gravel.

Materials

Below-ground mini-pump or fountain pump
Tub
Chicken wire
Plastic pipe, such as a wastepipe
3 pieces of plastic tubing
Newspaper
Steel wire
Fine concrete
Tile adhesive
Mosaic
Tile grout

Tools

Mixing pail
Trowel
Metal spatula
Bowl for mixing adhesive and grout

Method

1. Measure the height of the pump, including the pipes, and the diameter of the tub, in order to make the fountain the right size.

2. Shape a mushroom, including the cap and stem, out of chicken wire. Thread the piece of plastic pipe through the middle of the mushroom, then fill the mushroom with balls of newspaper. Shape the bowl in the same way, fill it with newspaper and connected the two parts with pieces of steel wire. Place three small pieces of plastic tubing evenly throughout the bowl to allow water to flow through them.

3. Shape a layer of concrete on top of the mushroom's cap and over the bowl, pressing the concrete down in between the mesh of the chicken wire. Make sure that no concrete gets into any of the pipes. You may have to use several layers of concrete to completely cover the chicken wire. Allow the concrete to dry overnight and then turn everything over and cover the remainder of the mould in concrete. Allow the concrete to dry until the following day.

4. Dilute the tile adhesive with water to form a gluey mass (refer to the instructions on the package) and then spread the adhesive on top of the concrete using a metal spatula. Place the mosaic tiles on the tile adhesive, working on a small surface at a time so you can experiment with different colours and patterns. Allow the adhesive to dry overnight and then grout with tile grout (read the instructions on the package).

5. Place the pump in the middle of the tub. You may like to bury the tub so it doesn't stick up above the ground. Place the fountain over the tub so it rests on the edges of the tub. Thread the pump's pipes through the plastic

tubes in the middle of the mushroom. Fill the tub with water (refer to the instructions on the pump's packaging), and plug the pump in to allow water to pump out over the tiles.

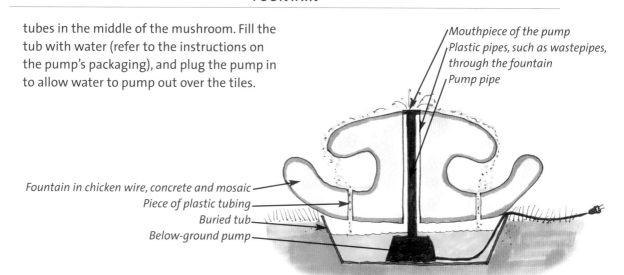

Mouthpiece of the pump
Plastic pipes, such as wastepipes, through the fountain
Pump pipe

Fountain in chicken wire, concrete and mosaic
Piece of plastic tubing
Buried tub
Below-ground pump

Leftovers

ometimes it can be hard to know how much concrete to mix for a certain casting. If you, like me, don't like to count and measure, you often have concrete left over. I mix by instinct and if the first batch isn't enough, I make a new one. Thus, out of the second batch, there is almost always excess concrete. So I take advantage of it by casting a small, simple object, like a pot or a candlestick, or I fill a plaster or soap mould from the craft store.

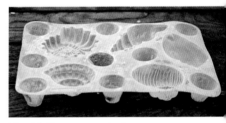

Moulds for soap or plaster are great for concrete, too.

When I mix concrete for large objects, I always make sure I have some small moulds ready. Plaster and soap moulds from the craft store require minimum preparation – they just need to be as clean as possible. You can even grab soft silicone muffin tins, beach toys and plastic boxes in different shapes to use as moulds.

Concrete shells cast in soap moulds. The shells were made from various types of concrete, which is why they are different shades.

141

Make sure that the moulds are clean and not full of leftover concrete from the previous castings. You don't need to oil them and can just pour in the concrete. Use an old tablespoon and fill them all the way up. Hold the corner of the mould and shake so that the air bubbles start to rise to the surface. Continue for a few minutes so you get less porous concrete surfaces. Cover with plastic and let your moulded 'soaps' harden until the next day.

Turn the mould over and press out the concrete shapes. Break off any thin concrete bits that stick out and smooth any rough edges using sandpaper. Wash the mould so that it is clean for the next time you have excess concrete.

Small concrete hearts, shells and roses make sturdy decorations for pots and flowerbeds. They can also be glued to picture frames, mirrors and walls.

The hearts below were cast in white cement and crushed marble with marine-blue pigment. For more about white cement, see page 37. For the roses to the right, I mixed oxide red pigment into the fine concrete.

Hearts in blue-pigmented white cement, cast in soap moulds.

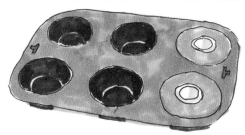

Create simple candle holders by pouring leftover cement into silicone muffin tins and adding a candle support from the craft store. Or cast a clip on a steel wire in a little concrete so it stands straight and can be used to hold photographs or memos.

Roses in red-pigmented cement cast in plaster moulds.

Compost bin

A compost bin doesn't have to be something ugly that's hidden away. It can, in fact, be a decoration for the garden. This double compost bin was built from light clinker brick, which is also called leca or mortar bricks. The blocks are made out of small ceramic pellets – leca pellets, mixed with concrete. They are available in different sizes from building materials stores.

The compost bin was mortared together – the pieces were fitted together like overlapping bricks. The surfaces were then plastered but not painted. A mortared compost bin will last year in, year out and it requires minimal maintenance.

Large compost bin made from plastered bricks.

Material

14 slender leca or other bricks, 59 cm (24 in) long, 29 cm (12 in) wide and 7 cm (2¾ in) thick

Mortar cement

Sand or macadam

Tools

Spade

Hard metal-strengthened saw

Hose with showerhead

Mixing pail

Trowel

Spirit level

Soft brush or sponge

Instructions

1. Dig a hole as large as the compost bin will be and about 30 cm (12 in) deep. Fill it with sand up to ground-level and stamp it flat. Use the spirit level to check that the top is horizontal.

2. Saw two of the bricks to make 4 pieces that are about 7 x 7 x 19 cm (2¾ x 2¾ x 7½ in) and 4 pieces that are 7 x 14 x 19 cm (2¾ x 5½ x 7½ in).

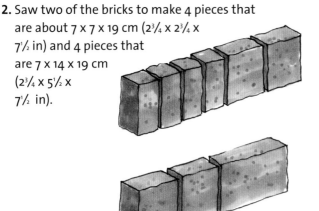

3. Lay the first layer of bricks following the formation in the diagram below. Mix the mortar cement according to the instructions on the packaging. Place a layer of mortar on to the bricks and press the second group on top. Scrape away any mortar that comes out between the joints. Continue in the same way alternating the two groups, placing mortar in the vertical joints. Use the spirit level to check the walls are level.

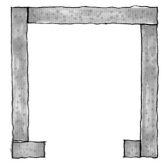

First and third layer of bricks.

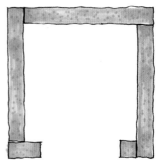

Second and fourth layer of bricks.

Plastering

First water the entire compost bin so all the surfaces are moist. Mix the plaster according to the instructions on the packaging and put a 1 cm (⅜ in) thick layer onto the bricks (masons use trowels to throw plaster onto bricks, as it sticks more easily this way). Smooth out the surface first using the wide part of the trowel and then with a soft brush or wrung-out sponge. The plaster needs to dry slowly so water it carefully with a thin spray a few times a day during the first few days or cover the entire compost bin with plastic.

Suggestions for gates

Attach small bolts on the inside of the compost compartment, as in the picture on the right. It is simplest to hammer to bolts in place, which works fine with light clinker bricks, but you can also use a drill and screws. Saw wooden boards that can slide between the bolts and the wall – use wood remnants treated with wood oil or buy waterproof timber. The boards should not be fixed together but should be taken out one by one when the compost bin needs to be emptied.

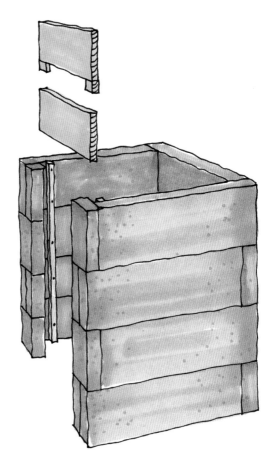

Kitchen sink

As a material concrete has made a real journey, from tunnels and bridges to architect-designed home decorations. Now concrete isn't just for facades and bridges but can also be used on interiors, for the floor, walls and furniture.

The best thing about concrete is that you can mix and cast it yourself and decide how it will look. We made a 6-metre (20-ft) long kitchen sink and work surface with many compartments and holes for herb pots and a sink. We made the holes with formwork plywood 'boxes' and pieces of cell plastic. The sink was cast upside-down against a piece of formwork plywood. It took a long time to make the mould and there were a lot of adjustments to be made before it was finished, but the sight of the new sink in its place in the kitchen was well worth all the work. We mixed concrete indoors in a large pail with the mixer on a drilling machine.

The following pages show the principles for a smaller and slightly simpler surface with space for a sink. Areas for a stove, enclosed shelves and flower pots are just a few examples of what you can come up with if you cast your own work surface.

The mould with spaces made out of formwork plywood.

Compartments for a plate rack, flowers and more in the finished sink. Photographed by Marco Pusterla.

Mould

Formwork plywood
Bolts
Lubricant spray or mould oil
Cell plastic
Silicone, uncoloured
Glue suitable for use with cell plastic
Wood glue

Materials

Fine or another concrete (we used
  Bemix standard), 5 sacks
Steel rods, 5 mm in diameter: 2 x 13 mm
  (½ in) long, 1 x 75 cm (29½ in) long and
  3 x 56 cm (22 in) long
Wire
Plastic

Tools

Ruler
Saw
Drill
Screws and screwdriver
Square rule
Cement mixer
Fine sandpaper
Large bucket
Trowel
Vacuum or rag

Before you begin make a simple sketch of how the sink will look, including all the measurements – width, length, thickness and the positions of the holes. Ensure the hutch can bear the sink's weight. We built a simple but sturdy table of bricks to place the mould on, but you can also put it on the floor, preferably close to where the sink will be positioned as it will be heavy to carry. Don't forget to protect the floor with a plastic sheet or thick paper.

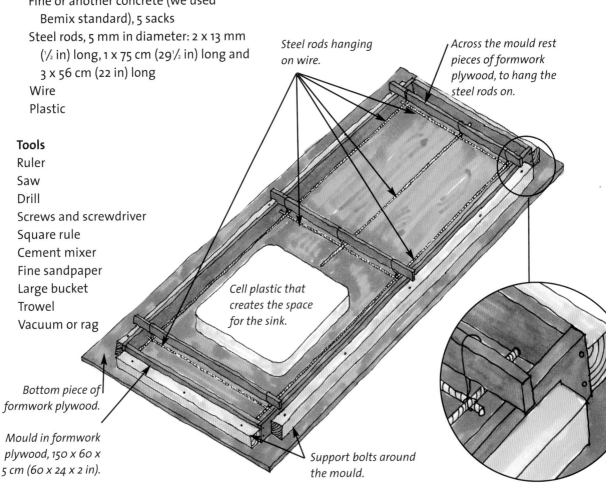

Steel rods hanging on wire.

Across the mould rest pieces of formwork plywood, to hang the steel rods on.

Cell plastic that creates the space for the sink.

Bottom piece of formwork plywood.

Mould in formwork plywood, 150 x 60 x 5 cm (60 x 24 x 2 in).

Support bolts around the mould.

Screw the wood for the mould together in the corners and then glue or screw the mould together using support bolts.

Fill in any cracks and scratches with wood filler and smooth them out.

Egil and Marco smooth out any unevenness in the wooden mould.

Marco vacuums the mould thoroughly before the concrete is poured in.

Instructions

1. Measure and saw the four pieces (the long sides and the short sides) for the mould. The height of the mould will be the thickness of the sink. Place the pieces on the bottom board and make sure you have measured and sawed correctly. Screw the pieces together in the corners.

2. Place bolts as supports and use the square rule to ensure the corners are accurate before screwing the bolts to the bottom board.

3. Glue the mould to the support bolts and leave it to dry. Smooth out any unevenness and/or fill in larger scratches with wood filler. Place silicone on the bottom board and in the corner joints and smooth out the silicone grout with your fingertips.

4. Carefully read the producer's instructions for setting in the sink. There is usually information about how big the hole (the sink measurements) should be. Cut out a piece of cell plastic to match the hole's measurements and glue it to the bottom board (the cell plastic creates the hole in the concrete). Remember that the piece of cell plastic has to be at least as thick as the final sink, in this case 5 cm (2 in). Tighten the joints with silicone so concrete cannot run out.

5. Time for reinforcement. Place sticks of leftover formwork plywood across the mould. Hang the steel rods on the sticks with wire. The steel rods should be in the middle of the concrete, in the middle of the sink's thickness, about 2.5 cm (1 in) up from the bottom board and not closer to the mould's walls than 2 cm (¾ in).

6. Vacuum the mould well or clean it with a damp rag – it is easiest to lift out the reinforcement at this point – then spray lubricant around the entire mould, or spread on a thin layer of mould oil.

7. Mix the concrete according to the instructions on the package. Measure the water carefully. Pour the concrete into the form when it is ready. It is best if there are two of you, so you can take turns. When the mould is filled halfway, you should press the trowel into the concrete so it ripples a bit to let the air bubbles come up to the surface so the top's surface will be less porous. Make sure the reinforcement isn't dislodged. Do the same thing when the mould is completely full. Smooth out the surface with the trowel. Place plastic tightly over the entire mould and leave it to harden for a day.

8. Lift the plastic and pour water over to cover the concrete. Wait a day. Add more water as the water evaporates. Do the same thing the next day.

9. After four days, you can loosen the sides of the mould and remove the cell plastic. Lift the piece into its place. Concrete is sensitive to fat and dirt, so treat it with stone soap.

Place sticks of formwork plywood or similar across the mould. Hang the steel rod pieces with wire so that they are about 2.5 cm (1 in) up from the bottom of the mould.

The first batch of concrete is poured in.

Tight joints in the mould create attractive and precise edges.

Further reading

- *A Garden Gallery* by George Little and David Lewis (Timber Press, 2008)
- *Creative Concrete Ornaments for the Garden* by Sherri Warner Hunter (Lark Books, 2005)
- *Making Concrete Garden Ornaments* by Sherri Warner Hunter (Lark Books, 2002)
- *Concrete as a Hobby* by Malena Skote (Prisma, 2006)

Book in concrete with wire spine by Bente Resaland.

Useful addresses

UK
The Concrete Society
Riverside House
4 Meadows Business Park
Station Approach
Blackwater, Camberley
Surrey, GU17 9AB
Tel: 01276 607140
Web: www.concrete.org.uk

US
American Concrete Institute
38800 Country Club Dr.
Farmington Hills, MI 48331
Tel: 248 848 3700
Web: www.concrete.org

Australia
Concrete Institute of Australia
PO Box 3157
RHODES NSW 2138
Tel: 02 9736 2955
E-mail: admin@concreteinstitute.com.au
Web: www.concreteinstitute.com.au

New Zealand
New Zealand Concrete Society Incorporated
PO Box 12, Beachlands
Auckland, New Zealand
Tel: 64 9 536 5410
E-mail: concrete@bluepacificevents.com
Web: www.concretesociety.org.nz

South Africa
Concrete Society of Southern Africa
PO Box 75364
Lynnwood Ridge, 0040
Tel: 27 12 348 5305
E-mail: admin@concretesociety.co.za
Web: www.concretesociety.co.za

Index